Industrial Studies Series

MANUFACTURING INDUSTRY SINCE 1870

Margaret Ackrill

Philip Allan

First published 1987 by

PHILIP ALLAN PUBLISHERS LIMITED
MARKET PLACE
DEDDINGTON
OXFORD OX5 4SE

British Library Cataloguing in Publication Data

Ackrill, Margaret
Manufacturing industry since 1870.—
(Industrial studies series).
1. Great Britain—Industries—History
I. Title II. Series
338.4'767'0941 HC254.5

ISBN 0-86003-528-X
ISBN 0-86003-630-8 Pbk

Typeset by Dataset, Oxford
Printed and bound in Great Britain by The Camelot Press plc, Southampton

Contents

Explanation and Acknowledgements

This book is planned on a grid. It may be read as a whole, or each of its four sections read as a self-contained unit. It is also possible to select a chapter, or a subheading within it, in any section — 1870–1914, 1914–1939, 1939–1951, and 1951–1980 — and follow it through the corresponding pages of the other sections. Each section has three chapters. Chapters 1, 4, 7, and 10 begin with a perspective on the period in question, outline the major discussions about it and then analytically describe the main features and factors of the British economy during it. These chapters are of importance in themselves and enrich the understanding of the chapters which succeed them. Chapters 2, 5, 8, and 11 discuss individual industries and the managerial choices in them, and chapters 3, 6, 9, and 12 deal with industrial relations and the laws, conventions and trade union organisation affecting them. The major topic not represented by a subheading in every corresponding chapter is that of manufacturing and research, about which there is more to say at some times than at others; and a few industries which illustrate an important point appear only at the time in which they had special significance. The aircraft industry does not have its separate entry until 1914–1939, and of course electronics makes an appearance gradually. With these fairly logical exceptions, the grid pattern holds, inviting the reader to see what happened before or later in any topic which creates interest, or to make comparisons between different epochs in British economic development and manufacture or between the fortunes of different industries or trade unions. The final pages provide a brief summing-up.

Apart from GDP and GNP, economic terms are explained as they occur. Full titles are given to organisations on their first mention, but they are then referred to by their initials: thus the Prices and Incomes Board becomes the PIB. The Index lists the full titles, followed by their initials in brackets. Appendix I assembles figures which are divided between the sections.

Explanation and Acknowledgements

It is not possible in a book of this nature to include references to the sources, primary and secondary, used in its compilation, but I thank and beg the indulgence of colleagues and other authors whose work I have used, as well as of those whose studies had to be ignored for reasons of space. I should particularly like to thank David Edgerton for his very valuable input into an earlier version of this book, and Mrs Sally Grosvenor and Mrs Elsie Hinkes for their clear typing of my untidy script. Thanks must also go to Professor Pickering, the series editor, and to the publisher, Philip Allan, and his assistant Mrs Helen Ramsay. Finally I should like to thank my family and friends for reading and commenting upon diverse pages.

The selection of topics of course implies a view of what is important for understanding manufacturing industries and their performance, but I have tried to be as objective as possible in the presentation of material, and to base conclusions, for which I alone am responsible, upon the evidence presented.

Margaret Ackrill
May 1987

Section A

1870–1914

CHARLIE CASH
BOILERSMITH'S LABOURER.

TOM TOYNE
BOILERSMITH'S LABOURER.

GEORGE MARKHAM
BOILERSMITH.

BILL WALTON
FITTER'S LABOURER.

CHARLIE DALLAS
FOREMAN FITTER.

GEORGE DIXON
APPRENTICE FITTER.

THOMAS SPENCLEY
PIPE FITTER.

BILL BARRON
PIPE FITTER.

JACKIE FOST
ROUGH FITTER

GEORGE ROBINSON
LOCO FITTER

TAKEN 1900.

The 'A' Team taken at Appleby Frodingham Steelworks, Scunthorpe, in 1900
Reproduced by kind permission of the British Steel Corporation, East Midlands Regional Records Centre

1

British Manufacturers and the Economic Environment

Of all the countries in the world in 1870, Britain had the highest output of goods per head. The distribution of these goods was very uneven, and life was hard for most people; nevertheless, the average working man and his family had a real income which exceeded that of his counterparts in any other country. A higher percentage (some 32½ per cent) of Britain's workforce was in manufacturing than in any other sector of the economy. This was greater than in any other country, and about 40 per cent of all the internationally traded manufactured goods in the world were made in Britain. In 1880, these exports were more than two and a half times more valuable than those of Britain's exporting rival, Germany. All this was because Britain had been the first country to undergo an industrial revolution. To understand the position in 1870, and changes which took place after 1870, we need to know something about the nature and causes of that revolution, and why it happened in Britain before it happened in other countries.

Perspective: Why was Britain First?

By 'Britain' we mean England, Scotland and Wales. The changes which resulted in the economy of 1870 were called a 'revolution' in an analogy with the French revolution. It is perhaps a misnomer, since the changes of the industrial revolution were spread over a century and the word 'revolution' carries overtones of sudden and violent change. However, there *were* marked and not always gradual changes in the appearance of Britain and the lives of its inhabitants. Canals and railways cut through fields and cities, new buildings changed the fields on the outskirts of old towns into manufacturing districts surrounded by new terraced houses, and small villages grew into bustling cities. People had to be accustomed to work by the clock instead of by the tides and the climate.

We call the revolution 'industrial' because the changes took place in a mixture of industries including mining and transport. Manufacturing itself changed its nature. The verb 'to manufacture' derives from a mixture of French and medieval Latin words, and originally meant 'to make by hand'. A manufactured article was broadly distinguishable from a product of agriculture (which grew things), mining and quarrying (which dug them out of the ground), and the provision of services (which made available for another person something which already existed). According to one definition, 'factories' were places where manufacturing took place; according to another, they were buildings or depots in which merchants or factors, as they were commonly called in Scotland, housed their wares. By 1850 a factory was a place in which water or steam-power was used in the manufacture of goods; by 1870, apart from hand-processes in the clothing and allied trades, 'manufactured' articles were those made with the use of machines, most of them power-driven. Skill and strength of hand were important, but the great increase in production, especially in the textile industries, came from power. Each of the causes which brought about this industrial revolution in Britain could find a parallel elsewhere. But the British combination was unique, otherwise Britain would not have been first to undergo this change. The main ingredients in the combination will now be examined.

Natural Resources: Water, Coal, and Iron

Britain had plentiful supplies of fresh water, essential for bleaching and dyeing textiles, and for sluicing and cooling in metal manufacture. There were streams and rivers to harness to provide water power; then the combination of increased demand for power and the development of the steam-engine established steam power. There was enough water to be diverted into the new canals. Canals reduced inland transport costs by half; then, as they were rivalled, and later exceeded, by railways in the amount of freight carried, there was enough water readily available for steam-engines, both locomotive and stationary.

Coal deposits were large and widely dispersed. The knowledge of how to extract coal from deeper mines came step by step as shallower seams were worked out. Knowledge of how to deal with coal's sulphurous fumes was also cumulative. Coal was used by the well-to-do for domestic heating in the sixteenth and seventeenth centuries, as their ornamental chimneys show, and for salt, brick, and glass-making. In the eighteenth century, it was burned in the first steam-engines which pumped water out of deep mines. In the nineteenth, coal fuelled factory, railway, and maritime steam-engines.

Iron ore was found in Wales and Scotland, and in some English counties. Swedish bar iron made the best steel for some specialised requirements, but British iron ore was acceptable in quantity and quality for most purposes until almost 1870. Coal was used to smelt iron ore in the eighteenth century, and in the nineteenth to convert iron into steel. Iron made the hinges and cross-bars of canal lock gates and turning gear, and railway lines, axles, and locomotives. Anchors, buckets, buttons, latches, gun barrels, pots and pans, textile machinery and winding gear for coal-mines were all made of iron. Steel, first reserved for knives, swords and some tools, was, by 1870, with cheaper processes, providing replacement railway lines, ships' plates, and strong girders.

Brick-clay and Timber

Stone supplied much building material in Scotland, Wales, and England's northern and western counties, but the presence of brick-clay to make bricks, and the development of cheaper canal transport to move them, was an undoubted asset. Canal bridges and cuttings themselves, and factories, houses and railway tunnels, were not held up for lack of building materials. Timber was a little more problematical. Hardwood forests declined by the eighteenth century, and softwood was frequently imported. Nevertheless, for local purposes, much timber could be found locally, diminishing, but useful.

Agriculture and its Products

While luxuries — sugar, coffee, and tea — were imported, and occasional bad harvests led to grain imports, Britain could feed itself until the nineteenth century. It is not essential for an industrial nation to do this, since its manufactured goods can be traded abroad for food. But it helps at an early stage if imports are to be resisted, and yet a rising standard of living obtained for the population as a whole. Leather for harness and driving belts of new machinery — and, more importantly, wool for woollen textiles, the dominant British export until it was supplanted by cotton textiles — were typically British products. Moreover, while in absolute figures the number of people working on the land grew until 1851, it declined as a proportion of the working population. Yet agricultural output grew faster than their numbers. This was the outcome of the agricultural revolution: like the industrial revolution, a long process of accelerating output.

Population Growth

Was the increase in the British population (10.69 million in 1801, 26.16 million in 1871) cause or consequence of the industrial revolution? During those same years, the population of France, which possessed a smaller industrialised sector, grew less (28 million to 36 million). At the moment, the consensus seems to be that the faster British growth was the consequence of industrialisation. More and more factory owners, who hoped to (and did) entice workpeople, especially young people, from the countryside, offered higher wages in the growing towns. Having higher wages, people were prepared to marry younger. Younger wives (average age of brides at marriage was 27 years in 1700–1749, and 24.3 years in 1800–1849) tended to have one child more than their mothers and grandmothers and so, because of bigger families rather than a decline in the death rate, the population grew. Since child employment was, and to a degree always had been, available, there was little incentive to limit family size. It is possible that the increase in family size noted in some rural parishes was due to a reduction in animal-borne disease, as the agricultural revolution led to better animal care. The outcome was a sufficient labour force for the growing industrial sector. Immigrant Irish labour offset the emigration of English, Scots and Welsh.

Skills and Inventiveness

Like all European countries, and most countries of European settlement, Britain had had a system of apprenticeship. A sixteenth-century law said that boys and, to a lesser extent, girls, should learn a trade or craft by working for a number of years with a recognised master craftsman. Only after satisfying a group of masters with a 'masterpiece' could the young person set up on his own as a master, or be employed for a full adult wage. This rigid pattern was never universal in Britain. There were apprenticeships, but people learned by doing, and took general dexterity from one trade to another. Nor were they prevented from moving from place to place by a guild of masters. Men with some skill with machinery were lured from eighteenth-century Scotland to Wales with the inducement of extra wages. New skills could be developed and imparted. Some of these new skills then developed their own rigidities, but that is a later part of our story.

Many improvements in production, such as the better setting of a furnace, were made not by theoretical and mathematical calculations, but by practical men on the spot trying this and that until they found something that worked. British workmen were renowned abroad for their practical cleverness. Even important discoveries, like Abraham Darby's use of coal to smelt iron successfully in the first half of the eighteenth century were not tied to any real understanding of why the process worked. That it *did* work led to admiration for the practical man, rather than the theoretician — understandable, but a dangerous attitude as technologies became more sophisticated. Yet there was a nucleus of mathematically and scientifically trained men (often the product of Scottish universities) who made important discoveries and inventions. Up to 1850, there seemed to be no shortage of people with the requisite knowledge to impel the industrial revolution forwards.

Capital

Canals, railways, mines, and factories and their machinery all needed capital — that is, money for the long term. The ratio of capital to people employed was lower in manufacturing than in these other ventures. Apart from some Dutch finance, capital was raised in Britain; there was no borrowing from abroad. Sufficient money had accumulated; there were channels and structures to make it available to British enterprise.

The profits of agriculture found their way into the development of canals and mines. The profits of commerce — and, Britain being already a trading economy, these were concentrated in ports, particularly London, Bristol, Liverpool, and Glasgow — were more likely to go into manufacturing or into nineteenth-century railway building. Manufacturing businesses could often set up on a small scale with little capital; premises were rented rather than bought. Raw materials such as cotton, or malt and hops for brewing in large town breweries, were obtained on credit, sometimes from the very merchant who then sold the finished product. Then the credit was renewed and in effect served as capital. Country lawyers sent sums for investment to London to be lent out to the manufacturing districts; canal shares were sometimes put on sale in the locality through which the projected canal would pass. Banks lent sums on a short-term basis but, renewed over and over, they too became capital substitutes.

Manufacturing enterprises, until the end of the nineteenth century, very rarely raised money on the London or provincial stock exchanges. This was partly because much manufacturing could begin in a small way. Where large amounts of capital were needed at the outset to sink the mine, dig the canal, lay the railway, capital was raised at first through partnerships, but then through companies publicly quoted on the stock exchange. Here the public could buy stock (that is, make an investment with a fixed rate of return) but not always shares (which stood to gain extra dividends if the company prospered). The stock exchanges, with their surrounding firms of jobbers and stockbrokers and merchant bank advisers, were good if expensive at marshalling funds for such ventures in Britain — and abroad. New laws made it easier to obtain limited liability of partners and shareholders for debts if an enterprise went bankrupt; this encouraged investment. Banks had to obey stricter rules, but as a result were safer places to leave money in — or borrow from.

There were also non-public ways of raising capital. Friends and families borrowed and lent among themselves. Nonconformists, often related by marriage, borrowed and lent within their own circle on the strength of what they knew of a man's character and prospects. A great deal has been written about the Protestant work ethic and its relation to the industrial revolution, or even the origins of capitalism. Protestant Holland possessing commercial wealth but lacking iron and coal did not produce an industrial revolution; Catholic cotton manufacturers, an enclave in some Protestant districts of France, displayed many of the same qualities as their British Nonconformist counterparts. In England and Wales, Nonconformist groups, excluded from Parliament until 1828, excluded from taking degrees at ancient universities (with all their temptations) until 1856, and excluded by their religion from gambling, excessive drinking, and illicit amours, were the mainstay of many commercial dealings and many manufacturing ventures. Abraham Darby obtained his capital from a Quaker merchant family in Bristol.

Lastly, the aristocracy and gentry of eighteenth-century Britain were just a little more inclined than their European counterparts to save, and so accumulate capital. It was put into productive uses in agriculture, or invested on their own or their relatives' behalf. Even while their estates were burdened with mortgages, money was found to make 'settlements' for second sons, about-to-be-married daughters, and maiden aunts. The money, if it did not go into the ventures mentioned, often went into government loans. British governments did not repudiate their debts, so the money was not wasted by the investor; its receipt by the government, while entailing the raising of taxation to pay interest on the loans, meant far less taxation of earned income than if the loans had not been forthcoming.

Social Mobility

Britain, in the years 1770–1870, had a considerable degree of social mobility, in spite of the notorious snobbery of sections of British society. Serfdom (still legal in France until 1789 and in Germany until

1848) had long since disappeared in law and in fact. Sumptuary laws which forbad a merchant's wife to wear the same fur hat as a duchess had been forgotten. Poor boys like Dick Whittington, who really did become Sir Richard Whittington, three times Lord Mayor of London, could rise to riches and social status. This was exceptional; nevertheless it was possible. Josiah Wedgewood, of pottery fame, set up his family in a handsome country house with a park, a collection of well-made furniture (and pottery), gilt-framed oil paintings. His family married into the established gentry. Money, if allied with manners, opened many doors, and manufacturing could provide money.

For those without such grand ambitions or achievements, the clubs and churches of nineteenth-century Glasgow, Leeds, Manchester, and Birmingham provided social status as well as other rewards. Large villas on the outskirts of manufacturing cities, and an influence in local politics, were part of the inducements to and benefits of success, though to do any manual work in a factory or elsewhere was to risk losing that status. Overall, social as well as financial gains were an inducement to industry; no other country offered more.

The Domestic Market

Commercial men, the well-to-do in agriculture, professional men and army and navy officers in Britain were middle-income earners with money to buy new manufactured products. In the early nineteenth century, perhaps not more than 10 or 12 per cent of British families were in this favoured category. The price of some goods, most notably cotton textiles, fell as the equipment on which they were made was improved and the scale of their manufacture grew. This fall in price meant that a new range of customers could afford to purchase sheets and shirts; profits could still be made as manufacture expanded, and so manufacture's employees themselves became customers. Railway building in Britain preceded railway building abroad; lessons were learned in the domestic market where defects were comparatively easy to rectify, and then applied when British engineers, rails, locomotives, and labourers were sent abroad. The domestic market was more important than the export one except for one or two products (especially cotton goods). While no other countries were producing rival exports, the initial gains by new industries in a comparatively prosperous domestic economy were consolidated.

Export markets

Britain conducted an extensive overseas trade. In the eighteenth century, British merchant ships sailed to all known ports of the globe.

Wool and woollen cloth had for centuries been the principal British exports, but merchants were constantly alert for new British goods. These were exchanged for timber, bar iron and high-quality luxury goods of Europe, the tea, spices, and silk of the East, the wines, citrus and dried fruits of the Mediterranean, and the increasing quantity first of tobacco, then sugar, and then cotton, from across the Atlantic. Sadly, until the slave trade was forbidden to British ships in 1807, they also wanted knives, guns, cloth, and trinkets to exchange with African slave dealers in return for slaves. All this built up established trading routes and merchanting organisations which could be used when the volume of goods available for export grew. Other countries took time not only to industrialise, but also to expand trade.

A great deal, in the eighteenth century, was decided by force. Britain was at war for over a third of the years 1700–1800, and whatever the origin of the wars, the outcome was often a change of owner of some overseas possession. With ownership went the right to fix a customs duty or even a prohibition on the goods of any other country seeking to trade with that possession. It may be that ownership imposed more costs than it was worth, in a particular territory. In 1783, the United States of America established their independence of Britain. The result, to the surprise of many, was an *increase* in British trade with the erstwhile colonies (in spite of some new unfavourable tariffs), and the end of the cost of governing them. The British occupation of India seemed to have a momentum of its own. In the eighteenth century it was prized as a source of tea, revenue for the East India Company, and military promotion; in the nineteenth, the Company had to relinquish much of its hold, and India became a market for British goods. It might not have been essential to have foreign markets; it might have been necessary only to export enough to pay for sufficient imports of raw cotton (the essential import of the industrial revolution) from the United States to satisfy the British home market. But British growth, even on the most favourable reckoning, would have been somewhat slower had its agriculture been further developed to maintain self-sufficiency and had exports of manufactured goods been foregone. There is no doubt that it was valuable to a variety of British manufacturers to have a variety of markets, including European ones. A slackening in demand in one could be compensated for by a growth in demand in another.

The Effect of War on Britain

In the eighteenth and nineteenth centuries Britain, unlike all other European countries, had only brief fighting on its soil — the Jacobean risings of 1715 and 1745 — and no foreign invasions. No part of Britain suffered war damage. The removal of young men from the civilian economy of Britain during the Napoleonic Wars of 1791–1815 took away men who at a time of transition in agriculture and industry might well have been unemployed. Certainly, after their return to civilian life, unemployment within two years became a problem as never before. On the other hand, war gave a stimulus to the development of metal industries and shipbuilding. Wars temporarily restricted imports and exports but, increasingly, since Britannia ruled the waves, Britain suffered less in this respect than France, Spain, and Holland, its maritime rivals. Nineteenth-century naval supremacy made the British navy a kind of international police force, keeping lines of legitimate trade open to all comers. Piracy and, eventually, the ghastly slave trade from Africa to the Americas were ended, with help from the British navy.

Government and the Industrial Revolution

What direct effect did the government have on industrialisation? Until the 1820s it tried (in vain) to stop skilled workmen emigrating, taking their know-how to possible competitors. Until the 1840s, it forbad the export of textile machinery and other equipment. There was a strong patents law, which protected and rewarded investors, though it could be used to prevent improvements to known devices. While trade within the Empire was virtually free of tariffs, tariffs were devised to keep out foreign goods (as well as to raise revenue) and British navigation laws insisted that goods be carried in British or colonial ships. These two sets of laws came under increasing attack; arguments against them were supplied by Adam Smith in his *Wealth of Nations* (1776). His basic theme was that a nation's wealth came not from the possession of gold, but from the enlargement of its trade. Trade and industry were most likely to flourish for the benefit of all if they were unfettered by government, even if a government's intention was benevolent, said Smith. The 'invisible hand' of the market would allocate resources, responding to what customers and consumers wanted, with the most efficient producer being the most successful. Competition, not protection, was the spur to wealth-creation. 'Free trade' described the absence of tariffs, quotas, and subsidies.

This doctrine became the basis of *laissez faire*, that is, leaving the economy and society to make their own way with minimal government intervention. By 1852, at least in the case of the British economy, it had gained acceptance in all major political parties and groupings. Protective tariffs were swept away and so, too, were the last of the navigation laws. While there were one or two very temporary compensating subsidies to agriculture, there were none at all for manufacturing; its expansion was unprecedented, and so seemed to prove the point. It had been argued that, by allowing the duty-free importation of foodstuffs into Britain, more would be sold, and thus the foreign producers of these foodstuffs, having an enlarged income, would buy British manufactured goods, and that indeed is what happened. Even British agriculture suffered no immediate blow, since it took time to establish new patterns, and in any case the growing British population could now afford more perishable foodstuffs, which British agriculture could best provide.

But just as these overtly economic laws were withdrawn, new laws with an avowedly social intent were introduced. It was asked how a prosperous nation could allow the very long hours and dirty and dangerous working conditions of factory and mine to continue. Adult men were held to be capable of making a fair bargain with an employer. After 1825 they were permitted by law to band together in trade unions to make that bargain. But women and children were held to be defenceless, and in industry after industry from 1833 onwards, inquiries, new laws, and well-enforced regulations established shorter hours and cleaner and safer premises, which benefited men as well. Child labour was gradually curbed and then, over many years, made illegal except for limited hours: children had to go to school.

Did this impose a cost on industry? The answer seems to be that it did not. Hours fell from 70 to 56 in the nineteenth century, but output did not decline, because less tired workers made fewer mistakes, and there was time to service and repair machinery and equipment which would otherwise have broken down during a long shift. Further, until 1870 and beyond, the superior skills and efficiency of the average British worker outweighed the advantage of lower wages in other industrialising nations.

The years 1850–1870, while 'golden' in retrospect, had their ups and downs, with profits and gains in living standards growing spasmodically rather than steadily. Nevertheless, by 1870, the British economy was in a very envied position. Expansion, improvement, progress, were the key words. The structure of British manufacturing — its main features and its economic environment in the years 1870–1914 — will now be outlined; bear in mind that the one constant factor was change, and that there has been much criticism of British industry during these years.

Complaints, Criticisms, and Controversies

Searching for the origins of the troubles of British manufacturing industry in the second half of the twentieth century, critics have unearthed a number dating from the period 1870–1914. Not content with accepting fairly uncritically the term 'Great Depression' which contemporaries applied to the years 1873–1896, and their worry about increased importation of foreign manufactured goods — 7 per cent of total British imports in 1850, 15 per cent in 1870, 20 per cent in 1913 — they also shake heads over the export performance of British manufacturing industry. They apparently think it should have expanded to prevent other countries increasing their share of world trade. There is criticism of the export of British capital, which they think should have gone to British manufacturing industry (or to social overheads such as housing, or to be distributed as wages). There is criticism of banks and of the London and provincial stock exchanges for making it difficult for comparatively unknown manufacturing firms to float shares successfully. More charges are that British firms did not form powerful cartels, as did German firms which then presented a united front in domestic and export markets, nor merge to form large, well-structured corporations, like American ones. A lack of applied science and scientists is also alleged and, finally, an ethos which has been held responsible for all these shortcomings: the decline of the industrial spirit. Value and esteem were attached to the life-style of the landowning aristocracy, and perhaps to professional pursuits, but not to manufacture. Thereby incentives to innovate and improve it dwindled.

How well-founded was all this criticism? Comments and judgements on some of it are made below, but a summing up is reserved until Britain's major manufacturing industries and their performance have been surveyed in some detail in the next chapter. The main concern in this chapter is to provide some understanding of matters which affected all British manufacturing industries.

Features and Factors

In these years, British manufacturing industry increased its output by an average of 2 per cent per annum. Numbers in its workforce, however, grew at only 0.8 per cent per annum from 1873 to 1913. Thus labour productivity, that is, output per person, had increased. But more capital was used, and so there was a smaller increase in total factor productivity than in labour productivity. That is, the increase in output not accounted for by extra inputs of either labour or capital rose some 0.6 per cent per annum on average. Over the whole period, exports of manufactured goods doubled; and nationally, while one or two years saw unemployment rise to just over 10 per cent, it usually stood at 5 per cent or less. Real incomes grew, in spite of some stagnant years, and prices fell. On the face of it, it looks as though the economy was not performing badly.

Size of Firms

Not until the census of production of 1907 do we have a reasonable picture of the size of the largest firms in Britain, measured by the number of people whom they employed. Using that as a starting point, the figures for the six largest employers in manufacturing in Britain have been established, as set out in Table 1.1.

Table 1.1 *Largest Employers, 1900–1914*

Employers	*Number of employees*
Fine Cotton Spinners and Doublers	30,000 in 50 plants (1908)
Royal Naval Dockyards	25,000 in 5 plants (1907)
Sir W.G. Armstrong Whitworth	25,000 in 4 plants (1906)
Vickers, Sons, and Maxim	22,500 in 8 plants (1907)
Calico Printers	20,495 in 26 plants (1903)
Great Western Railway (railway workshops)	17,700 (1913)

Source: Shaw, C. (1984) 'The largest manufacturing employers of 1907', *Business History*, Vol. XXVI, No. 1, March.

Of the hundred largest employers in Britain, the one with the largest single plant was the GWR, with 13,000 people in its Swindon works. Among the hundred, 6,820 was the average size of the workforce, spread between six plants. Of the 98 whose details can be established, only 26 had head offices in London. The picture is one of sizable works, with strong local roots for management. Outside these hundred, firms were of every size and shape. Small firms, employing 10 workers or less predominated in number, but in actual numbers more people found employment in larger firms. This was especially true of the major export industries: economies of scale in the textile and iron and steel industries led to large plant size, employing 200 or more as a norm in the former, and 2,000 or more in the latter.

Mergers and Concentration in Manufacturing

In 1870, manufacturing was very competitive in Britain. No one firm dominated the market for its products. But the following decades saw a growth in the size of the largest firms, measured by their capital, and an increase in the share of the market held by the larger firms. This is termed concentration. Seeking more capital, about 550 firms in manufacturing and distribution sought to raise it on the London Stock Exchange; others were floated on provincial exchanges in the years 1885–1907. Many of these shares were not floated to extend the capital of an existing firm, but to fund a new entity made up of a combination of firms. When share prices began to rise, other firms were encouraged to make use of company promoters and the services of bankers and lawyers and accountants, and float their shares also. The economies of scale of the larger firm also applied to financing: it made sense for firms to join in floating one publicly quoted company instead of paying fees individually. Investors lost money when anticipated profits from the new entities did not materialise and some new ventures lacked management which could control and reorganise them. However, other firms were successful. GKN, Imperial Tobacco, Vickers, and Watney were all put together at this time. Some of the most profitable retained their family management.

The peak year for mergers in Britain and in the United States before 1914 was 1889, but there were many more mergers and with a higher capital value in the United States. Perhaps one reason for this difference was the greater use made by US firms of the apparatus of the twentieth-century office: typewriters, telephones, accounting machines. Successful restructuring in the USA was more frequent, and was then an incentive and an example to further mergers. It might also be important that, compared with US exchanges, the London Stock Exchange could more readily raise money for firms without merger and also that the development of company law in Britain led to four-fifths of registered joint stock companies remaining private, that is, they did not disclose their accounts, until 1914.

The outcome was, by 1914, the apparent domination by some of the largest firms of the domestic market for products as diverse as wallpaper and tobacco. Because of the capital now necessary to set up in some lines of business (unlike the small-scale beginnings of firms in the early stages of the industrial revolution) entry by newcomers into some products was difficult. But this was not true of all newer industries in which, after all, most growth might be hoped for; nor did dominance mean a monopoly in British markets, open as they were to the products of all nations. British governments, despite a variety of political associations and pressure groups, imposed no tariff barriers. Therefore, in spite of mergers and increasing concentration, manufacturing industry was still very much open to competition in Britain. Associations of manufacturers sometimes arranged markets and prices among themselves, but such associations were never effective for more than a year or two. The biggest mergers swallowed up a number of firms, and these caught the attention. The typical merger was a smaller affair, increasing concentration, but leaving room for domestic competition.

Location of Industry

Ironworks were established and expanded where large deposits of ore could be found, or near an extensive coal field, since a ratio of about three tons of coal to every ton of ore smelted was not uncommon in the years we are investigating. The building of the Cardiff docks by the Marquis of Bute was undertaken to provide for the shipment of the iron, steel and coal produced in South Wales. Ports were often at the mouth of rivers, where inland and overseas trade might meet. Bulky imports, such as tobacco leaf and raw sugar and cotton, were processed near their port of entry. Bristol was at first the dominant port for the English Atlantic trade, Glasgow for Scottish — hence their sugar and tobacco industries. But when Bristol became too congested, the cotton trade shifted to Liverpool, and cotton manufacturing grew up in its hinterland, its location reinforced by the growth of engineering firms which could supply — and repair — the machinery needed for its manufacture. There is a rational geographical explanation for such a development. It is also readily understandable that ships of larger draught needed deep water in which to be launched, and that this dictated the location of the larger shipyards.

However, not all industries were sited because of geographical advantages. Shipping and shipbuilding had used the Clyde for generations, but to cope with increased traffic and bigger ships extensive works and dredging were undertaken to preserve these income-earners in successive schemes, a triumph over geographical limitations. Skills and previous investment determined the continued location of the industry. A further factor was the presence of a local market for the goods produced; big cities such as Birmingham, if they had a nucleus of growing industries, tended to attract others of a different kind, to supply their inhabitants. The sheer size of London, growing as a port of entry from Europe, and as the seat of Parliament, the Government, the Law Courts, and fashionable society and its pleasures, led a diversity of manufacturing industries to its environs.

Nevertheless, it had high unemployment, and many casual workers.

Thus, in the years 1870–1914, apart from London and those portions of the Home Counties which served it, the most prosperous parts of Britain were the northern half of England, the Clyde valley of Scotland, and South (and a little of North) Wales. The staple industries, by which we mean coal-mining, iron and steel works, heavy engineering, shipbuilding, and most textile production, were situated there. All these industries had altered their methods in the industrial revolution. They provided together over 40 per cent of all jobs, something like the same proportion of the gross domestic product, and even in 1914 over 80 per cent of all exports or, if we subtract coal, over 70 per cent. The fortunes of many of the industries were interlocked. Demand in shipbuilding fed back into demand for steel, iron, and coal. The wellbeing of these regions, with their predominantly low rates of unemployment, was tied to the dominant staple industries.

Population, the Manufacturing Workforce, Employment, and Education

The population grew as follows:

1871	26.07 million
1881	29.71
1891	33.03
1901	37.00
1911	40.83

Whether, in our interest in manufacturing industry, we look at population as a potential workforce, or a potential market, or a potential burden which has to be carried in part by the good performance of manufacturing industry, this is a growth to be reckoned with. The biggest increase, both absolutely and relatively, was between 1891 and 1901. Its main cause was the decline in the death rate of men and women in the very employable 30–34-year age group.

Growth would have been even faster but for the decline in the fertility rate, that is, the number of babies born to women of child-bearing age. The fall was most noticeable among occupational groups which tended to have smaller families anyway. Deliberate restriction of family size (although accompanied by the survival of more babies) led to a decline of 33 per cent in family size for British professional and higher administrator fathers; 21, 20, and 15 per cent for skilled, semi-skilled, and unskilled fathers respectively; and miners' and agricultural labourers' families declined by 10 and 15 per cent respectively. This means that these groups from which men in steel works and factories were drawn (two-thirds of the sons of skilled men followed their fathers into the same skills) were more than keeping up their proportion in the economy. More jobs would be needed for them in due course: in 1913, 32 per cent of the working population was employed in manufacturing.

Married women were less likely to work in 1911 than in 1871; by 1911, less than 10 per cent were in employment. The growth of heavy industry, steelworks and shipyards, meant that there were areas with job opportunities for men, including comparatively well-paid and steady jobs, which nevertheless did not offer many opportunities for women. In any case, in many communities, such as the coal-mining villages of the Welsh valleys, it was a point of family pride that a wife did not 'go out to work'. Domestic service, including both full-time and part-time work, was the largest single category of female employment. Many men and their wives could maintain a better standard of comfort by the wife devoting her time to cleaning and cooking, making-do and mending within the household than by her accepting this part-time, low-paid work outside it.

The introduction of old-age pensions in 1908 for men and women over the age of 70 (provided they had not received poor relief and did not have an income exceeding £13.50 per annum) probably led to some reduction in the participation rate, that is, the percentage of people wanting to work. Schooling withdrew hitherto unschooled children from work, reducing the participation rate further.

Nevertheless, there was no sign of a shortage of numbers in manufacturing. Had there been, there was available a considerable source of labour from the shrinking agricultural sector. But these people, to judge from the statistical record, went into the service sector. By 1913, workers in transport, communications and commerce outnumbered the manufacturing workforce. A good deal of this reflected the increase in food processing and its transport and retailing through chains of shops. If we add public and professional services to transport, communication and commerce, all categories within the wide definition of the service sector, we find over half of Britain's working population so employed: the rise of the service sector is not a recent phenomenon.

Education extended under the 1870 Education Act was not designed with the needs of the manufacturing industry in mind, in spite of the rather fanciful interpretations of some educational sociologists. Indeed, manufacturers in many instances opposed compulsory schooling when it arrived in 1876. Before widespread free education was available, factory employees, especially in Manchester and surrounding towns, had voluntarily paid small weekly sums to have their children taught. By learning to read and write and do basic arithmetic, they could find jobs as store-keepers and clerks, away from the danger and dirt of actually tending machinery. Over 80 per cent of children spent some years in school *before* 1876. The overt aim of the government in introducing the 1870 and subsequent education acts had been political — to enable boys, who as men would be given the vote, to read and write. This, it was hoped, would prepare them against the rash spoken word of politicians. Girls would grow to be better wives and mothers if they, too, had these accomplishments.

Yet the disciplines of regular attendance and the need to concentrate were a useful preparation for any employment (these features of education are sometimes described, rather sinisterly, as part of 'social control'). Also, the arithmetic learnt was useful for counting and measuring; basic skills were necessary for a variety of semi-skilled work, and also for apprenticeships in engineering and carpentry. Elementary schools before 1900 provided more advanced lessons for their older, abler pupils, and instructed some of them in basic scientific principles, and in technical drawing. 'Central' schools were then established for older pupils: their emphasis was not scientific or technical.

In most public (i.e. fee-paying and boarding) schools, science, if taught at all, was not well taught: in other fee-paying schools it was often lacking, though Scottish schools were probably more forward-looking than others. Grammar schools, revivifed in the 1900s and charging small fees except to scholarship boys and girls, tried too often to imitate the curriculum and achievements of public schools. Grants were available to establish secondary education with a science bias. But only three such schools had been established by 1885, though these schools laid the foundations for the science departments in the grammar schools.

Rather more was done in an assortment of technical colleges, which had local government finance. In 1878 the City and Guilds of London Institute began, and in turn it established the Finsbury Technical College and the Central Technical College (this last became part of London's Imperial College). Enrolments, however, were often disappointingly low. In the 1890s local authorities set up colleges on the same lines, financed by means set out in the 1889 Technical Instruction Act. The emphasis was on theoretical understanding and application of known scientific principles; with some exceptions, these colleges were not in the forefront of research, though they did provide training in scientific testing and experimental methods for limited numbers.

Science degrees, established in the 1870s, did not have much prestige in Oxford or Cambridge, in spite of an excellent mathematical tradition in the latter. Applied science was further still from academic respectability and had no place in those universities. Scottish universities, some of the London colleges which combined to form the University of London, and some of the new provincial colleges which subsequently became universities had science and applied science courses but comparatively few takers. The Royal College of Chemistry was set up by Prince Albert, aware of the gap between Britain and his own country, Germany, where men trained in universities moved into industry, especially into chemical firms, and kept their links with current research. The College was absorbed into the Royal School of Mines. The latter was a small institution, with at least half its student intake coming from overseas.

Managers were usually the owners or family of the owners of enterprises. Literacy and numeracy could be assumed, and some had at least a smattering of accountancy or legal training, but before 1914, apart from a small start in Birmingham, management training scarcely existed. Engineers read and circulated papers on applied research to their Institutes, but had no financial or managerial training. They were, however, given senior management posts when family circles could not provide adequate managers.

Thus British education before 1914 was not geared to the needs of newer, science-based industries, nor to the control of the larger new corporations. The values of the men and women who taught young people, apart from some dedicated specialists, were likely to embody an escape from, rather than an improvement in, British manufacturing. Too often, new, industry-related courses were not taken up when they *were* established, and employers continued to be hesitant in employing theoretically trained people. 'Practical men' were preferred.

Capital

In the Britain of 1870 to 1914, the propensity to save was quite high compared with the 1920s and 1930s, though perhaps a little lower than in the 1850s and 1860s. Did enough of those savings get directed into British manufacturing industry to ensure its wellbeing in the medium and long term? If not, why not?

Demand for capital from any sector of the economy is bound to change. At a time of little technical innovation but an expanding population, and markets which expand proportionately, we would expect to find fixed capital increasing roughly in proportion to population increase. Capital input and total factor productivity did increase, yet was it enough to keep the competitive advantage which Britain undoubtedly had in 1870?

Many decisions to rely on well-developed skills rather than to put capital into new types of machinery which were a substitute for those skills were sound. Examples will be found in the next chapter. The crucial test was the unit cost, the cost of a finished product, not the absolute cost of labour, nor of machinery and equipment, nor the cost of raising capital. On the world market, from 1870 to 1914, the cost and selling price of British goods was no higher on average than that of competing goods from other countries, unless there was deliberate dumping, that is, selling below profit or even cost. There *were* technical and product changes which used capital but there was also an attitude of caution about capital expenditure, the more understandable by the reluctance of trade unions to agree to change (see Chapter 3).

There were three disadvantages to this caution, however. The first was that making incremental improvements to fixed capital made it more financially disadvantageous to scrap it altogether, as one might be loth to move from a house which no longer suited one's requirements, but on which one had spent sums on alterations and improvements. The second was that reasonable caution might breed an attitude towards capital expenditure, and/or the idea of completely remodelling a factory or a shipyard, which prevented expenditure when it really *was* desirable. The third was that any increase in production was likely to be slow with existing equipment, yet markets could expand rapidly.

Had manufacturing industry asked for more capital, would it in fact have been available within Britain? Contemporaries had no doubt that it would, for three reasons. The first was that in 1914 and later, when the government called for loans to support war expenditure, there were large flows of funds in response to its appeal. The second was that, when large mergers called for funds through the London Stock Exchange in 1899, they were inundated with applications for shares. The third was that Britain annually invested a large and increasing amount of capital abroad. This capital could have been retained in Britain had manufacturing industry asked for it.

These arguments do point to a sufficiency of capital. But was there another range of reasons which prevented manufacturing industry from raising capital? Did the mechanisms of reinvesting profits, raising bank loans, and contemplating an approach to the Stock Exchange (in spite of the tenfold increase in manufacturing firms with a Stock Exchange quotation) have disadvantages and limitations not apparent earlier?

The answers cannot be clearcut, but we do have some general indications. *Reinvesting profits* was less good than it had been because, on average, profits fell in the years 1873–1896. There was not as much of them to invest. These were the years of the 'Great Depression', when total world trade grew less fast than it had done in the previous twenty years; and real and money prices of many commodities fell.

Profits in manufacturing industry were still a healthy 12–15 per cent, but this was less than the 25 per cent and more experienced previously.

In the years 1873–96, the rate of bankruptcy in particular localities increased. *Banks* therefore became more reluctant to lend to manufacturing industry. And, as smaller banks merged with bigger ones, less discretion was allowed to local managers, and it became more difficult for firms to acquire and keep bank loans. At a time when German banks were becoming more closely allied with manufacturing industry in Germany, British banks were distancing themselves from it. Banks had virtually no industrial expertise.

The *Stock Exchange* had its perils: instances can be found of firms not raising more than a fraction of the hoped-for capital, yet having to meet all the fees, including those of advisers who did not give them adequate publicity. Such instances discouraged other firms from trying.

Yet when the limitations of all three methods of capital-raising have been noted, the fact remains that firms which persisted in their efforts were able to overcome these obstacles and find capital. The structure of British finance may have been a hindrance, but it was not a barrier to the determined, as Chapter 2 will demonstrate.

Export of Capital

Capital was going abroad not because the rates of return were higher than in British manufacturing industry — they were usually considerably less. But they were slightly higher than on the same categories of investment in Britain. These categories were railways, municipally-owned harbour, water, and gas works, and the public works bonds of developing colonies. Their appeal was their security, for overwhelmingly they were in countries with good growth prospects and political stability. Investment in overseas subsidiaries of British companies, and personal investment, was scattered: of overseas stocks and shares bought in Britain, some were for Europe, but over 50 per cent went to North and South America; 14 per cent to Asia; 11 per cent to Australasia; and 11 per cent to Africa. Of this last, most went to South Africa, especially to its mines and the emerging infrastructure. Investment in British manufacture might well have been more risky financially.

Contemporary critics claimed that, instead of allocating so much to profits which were then sent abroad, industry should have paid higher wages to its workforce. This would have shown greater social justice, and at the same time increased the home market for the products of British manufacture by putting more purchasing power in the hands of those who would spend it in Britain. Yet exported capital helped develop the infrastructure of countries with climates suitable to grow the foodstuffs in demand in Britain. Their produce — wheat, meat, butter — kept down the cost of living for British working class families, many of whom could not afford the higher prices of some home produce. The capital itself was used to buy rails and girders, railway rolling stock and gas pipes, much of which came from Britain although, after the turn of the century, Germany and the United States obtained orders paid for by British capital. Yet this too added to world trade. Farmers and growers and miners and steelworkers abroad, having sold their products, bought goods from Britain. Fine English bone china, wreathed in rosebuds, could be found on the plains of Kansas. The benefits of capital exports were not confined to those who drew dividends from them.

A further argument was that the capital should have been retained to rebuild the squalid and congested housing in which many Britons lived. Slum clearance was only in its infancy. But had capital exports been restrained and the money put into housing, food would have cost more, and while there would have been a temporary increase in jobs in housebuilding, that proportion of manufacturing industry which was dependent on overseas demand would have declined and jobs fallen. In any case, a restraint on capital export, in the prevailing climate of free trade, was scarcely a political possibility.

Research seems to show that previous estimates of the amount of capital exports were exaggerated. The idea that in some years half of all British investment went overseas must be open to considerable doubt. Further, when British firms which already traded abroad set up subsidiaries overseas, they did this by using the profits of their trade as capital, not by sending capital from Britain. This adds to the contention that the export of capital has been wrongly blamed for alleged shortcomings in the British economy. It is difficult to quantify the benefits of capital export, since so much depends on counterfactual arguments, that is, suppositions about what might have happened had it not taken place. But it can be shown to possess some advantages to manufacturing industry, and was both cause and consequence of the growth of world trade.

The Domestic Market

Growth in the domestic market for British manufacturers depended on an increase in population and an increase in average purchasing power. Between 1870 and 1914, both enlarged. Increase in purchasing power came through a fall in the cost of many items: more could be bought for the same money. There were some stationary years, but the trend was upwards.

Some major railway, gas and water works were completed in these years. By 1891, useable railway mileage was at its peak. Increased shipbuilding, both merchant and naval, provided part of the alternative use for iron and steel. The coming of the railway meant wider markets. In Britain, local and regional differences in taste were greater than in some other countries. Where new technologies made large-scale production cheap and practicable and transport cost less, advertising was used more extensively to build up a wider range of customers. Mass circulation daily newspapers, themselves the product of new technology, were founded in the 1890s, and carried advertisements. So, too, did the end and side panels of horse- or, after 1900, electric-powered trams. After 1871, a series of poor harvests, and then the arrival of cheaply transported grain from the USA, undermined the profitability of farmers and landowners who clung to grain, and the wages of their poorly paid labourers. Their purchasing power was comparatively low.

Some British manufacturers became very concerned about increased imports of foreign manufactured goods, particularly German ones. A book, *Made in Germany*, appeared in 1896. It drew attention to the import of not only toys and watches, but semi-finished steel and engineering products which 'ought' to have been made in Britain; but no tariffs resulted.

There were fears of 'dumping', but there was little evidence of this except in iron and steel, although there was some price cutting of imported goods in Britain's free entry market. This, rather than lower wages elsewhere (since low wages were often accompanied by low labour productivity), was the cause of the low price of some foreign goods. The challenge to particular industries from imports is reviewed in the next chapter. The value of manufactured exports from Britain was by 1914 still more than three times greater than the value of manufactured imports into the country. The domestic market bought British goods in the main.

Export Markets

Not until 1871 did Germany assume the territorial shape which lasted until 1919, and the last pieces of the jigsaw which went to make Italy were also fitted in during 1871. The USA, recovering from its civil war of 1861–5, opened up new land for settlement in the Middle West, and new railways crossed the country from coast to coast. In Germany and the USA new-found reserves of coal and iron were developed, forming the basis of steel works and allied industries. Both countries were bound to have the advantage over Britain in delivering these goods to their own home markets, and to the markets of the countries immediately adjoining them. The USA and many European countries experimented with free trade in the years 1850–70; then, with the downturn in profitability and price levels of 1873–1896, strongly nationalist governments imposed new tariff barriers. By 1904, the tariffs against classes of British goods most commonly imported into Germany were 25 per cent; Italy imposed 27 per cent; France 34 per cent; the USA 73 per cent; and Russia 113 per cent. Even so, British imports entered these countries; but it is understandable that British manufacturers looked for markets outside Europe and the USA, in British colonies (though Canada, of course, turned to the USA for many goods), Argentina, and the colonies of other powers. British exports diversified, but most were of staple goods. About one third of British exports went to Europe still, but British salesmen were energetic enough to find alternative markets.

Naturally, the industrialising countries also looked for new markets. Their share of the world's traded goods increased, as their output grew faster than Britain's, in part because they were doing what Britain had already done, albeit in a shorter span of years, and in part because they were developing new goods. Table 1.2 compares the share of the United Kingdom with that of six other industrial countries in world exports of manufactured goods (it should be noted that the UK figure includes Ireland). These figures represent a challenge to, but not a defeat of, British exports which in *absolute* terms were growing.

Table 1.2 *World Exports, 1881–1913 (%)*

Year	*UK*	*US*	*Germany*	*Italy*	*France*	*Japan*	*Sweden*
1881–5	43.0	6.0	16.0	2.0	15.0	0.0	1.0
1899	34.5	12.1	16.6	3.8	14.9	1.6	0.9
1913	31.8	13.7	19.9	3.5	12.8	2.5	1.5

Source: Matthews, R.C.O., Feinstein, C.H. and Odling-Smee, J.C. (1982) *British Economic Growth, 1856–1973*, p. 435, The Clarendon Press.

The Balance of Payments

The figures for the balance of payments show whether Britain in any given year receives more or less income from abroad than it pays out overseas. They include payments on 'invisibles' — insurance, loan charges, shipping freight payments — as well as on goods and commodities. From 1870 to 1914, there was little concern about the balance of payments. While there was a deficit on goods and commodities, this was more than offset by the income derived from invisibles. London was the world centre for insurance, and Britain had by far the largest merchant fleet. Nevertheless, there was a rather unreasonable concern over Britain's deficit with individual countries. There was particular headshaking in the 1900s over a deficit with Belgium, the second country to undergo an industrial revolution. Since there were well-established ways of multilateral debt repayment, there was no real cause for alarm over any particular balance. Deficits were put right by credits with India and Australia. Manufactured goods from Britain, imported by those two countries, were major items in establishing an overall balance favourable to Britain. The income from capital invested abroad was also a useful source of income. Britain was the world's largest international creditor.

The Role of Sterling

Because Britain was a net creditor, and because so many of the world's international money payments passed through London, sterling balances were left in Britain, and particularly in London, and could be used for short-term loans. The value of sterling was not in doubt. It was readily given and received in transactions, reflecting the underlying wealth of the country, and was readily convertible into gold. Anyone presenting valid banknotes could insist on exchanging them for the same nominal value of gold sovereigns, but banks reckoned that not everyone would want this. Countries which had a partly silver base to their currency did not find their currency as much in demand.

In theory, the amount of gold in Britain was supposed to ebb and flow with external debit or credit. Should prices rise too much, international customers were presumed to withdraw from sterling, and gold would leave the country. Credit in the economy would then be restricted, prices would fall, since there would be less money to keep up demand, and then gold would come back to take advantage of low prices. In practice, things did not quite work out that way. The Bank of England was a private company, but it was the Government's banker. As such, it advised on and virtually set the rate of interest which the Government had to pay on its long-term debt, and it held a market, which it could influence, in short-term debt. But it also used the rate of interest to regulate the flow of credit — and gold — putting interest up perhaps ½ per cent when it saw signs of an ebb, putting it down when the inward flow was re-established again. These operations helped keep the value of sterling steady both in relation to gold and to the currencies of other countries, provided they did not devalue themselves. This stability had certain disadvantages, but they were not apparent in these years. The strength of sterling, partly created by the profitability of British manufacturing industry, was an asset, and something which was taken for granted.

Government and the Economy

After the victory of the Liberals in the 1906 election, a new attitude was shown by government towards intervention in the economy, with its accompanying increase in expenditure. Old-age pensions were followed by the National Insurance Act of 1911 which established two further sets of provisions. One set, for some 2¼ million workers in the construction, shipbuilding and engineering industries, where the demand for labour fluctuated with the cyclical state of trade, provided unemployment benefit for up to fifteen weeks a year. The money came from employers, employees, and the government; trade unions were one of the avenues through which it was paid out. The other set, covering some 13 million workers whose income was below a given standard, was also contributory, and provided sickness benefits. The first benefits from both schemes were paid out in 1913. An Act of 1909, based on the belief that much unemployment was simply the outcome of ignorance of what jobs were available, set up Labour Exchanges. This belief, in the circumstances of those years (particularly in view of the host of casual workers in London), had some justification. In 1913, a million jobs were found through the exchanges.

There were other reforming laws, such as school meals for undernourished children, and school medical inspection which helped improve the health of the future workforce. The 1888 Local Government Act tried to bring order to centuries of piecemeal and eccentric laws regulating local government. There followed road building and repairs, an asset to all firms which did not have their own railway-siding. Before 1914, most firms used horse-drawn drays for the local delivery of goods or their initial journey to a railway station. A few bold firms, particularly those involved in food processing, had their own motor delivery vans by 1914. This led to more pressure for better roads, and therefore more expenditure.

At the same time, government expenditure on defence also increased, for three reasons. The first was the low level of expenditure on defence over many years of peace. Weapons and equipment at last had to be modernised. The second was the Boer War of 1899–1901. The third was the naval programme of 1905, given justification by the big increase of the German fleet in what were traditionally British home waters.

The effect on manufacturing industry of these expenditures was twofold. First, there was an increase in orders for everything from battleships to bullets. (Defects in the organisation of the munitions industry, shown up by the Boer War, were not put right, however.) Second, there was an increase in taxation to pay for defence. By modern standards, this taxation was very light and there was no tax on companies. Income tax, down to a low of 2p in the £ in 1874, was doubled by 1885, and was roughly 7p after 1906. In 1907, tax on unearned income, which included dividends on shares in manufacturing, was charged at a higher rate than that on earned income. Only about 10 per cent of the population paid any income tax at all. About 50 per cent of national revenue came from indirect taxes, such as high excise duties on tobacco and alcoholic drinks. As a percentage of gross national product, taxation grew from 4.4 per cent in 1870–4 to a peak of 8.7 per cent between 1900 and 1904. By 1910 it was down to 7.4 per cent — not a heavy burden.

Of all government expenditure, about half from 1870 to 1914 was by local government. Of central government expenditure, the main item was defence. There was no thought of using government expenditure to affect the economy in any overall sense, though occasionally the Foreign Office tried some grandiose and usually fruitless scheme to enhance British influence abroad by initiating capital expenditure — provided not by government, but by company funds. The one successful venture resulted in the Anglo-Persian Oil Company, forerunner of BP. There were no subsidies to industry at home, and when help was given to chartered companies abroad (on a limited scale), it was because they served a political rather than an economic purpose.

The quality of help given by British consuls abroad to British exports varied considerably: there was no department of the Foreign Office with any effective control over economic intelligence, as it was called. Most British governments did not contain men with experience in manufacturing industry, and such people were scarce in Parliament, too. The most useful part of government to manufacturers was considered to be the information in the Board of Trade returns, which at least gave some broad statistical information about the nature and direction of traded goods and commodities. Manufacturers, far from being helped by the government, felt themselves to be not understood by it, and an energetic few, such as Dudley Docker, a powerful figure in the Midlands, tried in vain to rectify that through political organisation.

Conclusion

We have surveyed the factors involved in the industrial revolution, and the major factors and features affecting the manufacturing sector of the British economy in the years 1870–1914. Clearly, it was a sector under challenge, but still with considerable strength and vigour. In 1914, it provided employment for millions, and contributed to a standard of living unsurpassed except in the USA, with its vast natural resources. We shall now turn to individual industries to examine the possibilities and limitations in their responses to change.

Further Reading

Ashworth, W. (1960) *An Economic History of England, 1870–1939*, Methuen.

Floud, R. and McClosky, D. (eds) (1981) *The Economic History of Britain Since 1700*, Vol. II, Cambridge University Press.

Mathias, P. (1983) *The First Industrial Nation*, 2nd edn, Methuen.

Wiener, M.J. (1981) *English Culture and the Decline of the Industrial Spirit*, Cambridge University Press.

Questions

(1) What factors were important in leading to British dominance in world manufactured exports to 1914?

(2) Was the fall in the British share of internationally traded manufactured goods inevitable between 1870 and 1914?

(3) What features of the British economy encouraged or discouraged manufacturing industry between 1870 and 1914?

2

Old Industries, New Industries, and Managerial Choices

It is a commonplace to speak of old and new industries in Britain in the years 1870–1914. It looks simple enough. Old industries were those which were well-established in product and method and had for decades been considerable employers and large-scale providers of goods. New industries were those making new products, by methods which sprang from applied scientific research, rather than from incremental improvement in method.

On closer inspection, things are not quite so simple. 'Old' industries, such as iron and steel and shipbuilding, adopted quite new methods or produced a quite different type of product in the period; 'new' industries, such as the chemical industry with its scientific orientation, made products which at least for many years had been made by 'old' methods.

However, if the word 'old' is applied to staple industries and 'new' to almost everything else, with engineering bridging the gap between the two, we shall conform to the usual distinction. It is not possible to survey all British industries, so some which were of major importance to the British economy because of their size, their interest, or their portent for the future are selected.

In every industry there were important choices to be made. These choices were made by the top management of firms which consisted of paid managers, allowed considerable influence and discretion in the running of a firm, but rarely given a seat on the board, and of a firm's owner-founders, or their descendants. (Owner-managers are called entrepreneurs — that is, they are the link between their firms and the market. The term is more readily applied to energetic men; at times it has had disparaging overtones of sharp practice or dishonest dealings, and so is not used in this book.)

OLD INDUSTRIES

The Textile Industry

Cotton

This was the first industry to undergo great expansion of output and of productivity during the industrial revolution. Its net output in 1907 was £45 million and it exported about three quarters of that output. It employed about one tenth of the British manufacturing workforce in 1910 and so, by any standards, its fortunes were of importance to the British economy. In 1913, it provided 24 per cent of all British exports. All raw cotton was imported; imports mounted at a steady 1.1 per cent per annum between 1875 and 1900, and 1.6 per cent per annum from 1900 to 1912, an indication of rising sales. Most cotton came from the USA; Indian cotton never supplied more than 10 per cent and usually about 5 per cent of British needs. Egyptian cotton, which could be spun both fine and strong, was a small import, going to specialised firms. 83 per cent of cotton workers were employed in Lancashire and adjacent Cheshire.

In Britain, cotton was traditionally spun on spinning 'mules'. Their design had improved gradually. The six spindles of the first eighteenth-century mules soon increased to 60; by 1887 three skilled Lancashire spinners (almost all spinners were men) could tend 1,000 spindles. At about the same time, 5.8 men were needed to do the same work in Alsace, 6.2 in Switzerland, 7.2 in Saxony, and 16.6 in Russia. What is more, British spindles turned faster than their foreign counterparts and this increased productivity. Spinners benefited from higher wages and shorter hours than their foreign competitors — though not without a struggle with employers.

By 1913, perhaps a quarter of British spindles were the newer ring spindles, which could be tended by cheaper, semi-skilled labour. Would it have been an economic advantage if British managers had adopted more of this equipment? The idea that it would have been, and that it was the conservatism, even stupidity, of British mill-owners or the hostility of spinning trade unions which prevented it, has been the basis of much of the judgement that British industry had somehow 'failed' by 1914.

But there was a good case for retaining mules in British mills. Given the skill of the operatives, their lower wages than their USA counterparts, and the cost of installing ring spindles, it made economic sense to use mules for the fine and medium yarns which most of the British weaving industry wanted. Ring spindles, to produce yarn of a high count (that is, very fine) needed more expensive, longer-stapled cotton fibre than did well-managed mules. Further, ring spindles spun onto larger, heavier bobbins than did mules. Unlike the American cotton industry, the British was not vertically integrated, that is, most firms were one-stage firms, and so the transporting of heavier bobbins to weaving 'sheds' would have added appreciably to British costs. There would have been no difficulty in purchasing ring spindles in Britain had they been wanted: some of the best were made in Lancashire.

Similarly, the USA and other countries used fully automatic weaving machines — very expensive, and needing a different type of building. Initially, they were best at producing coarse cloth, which was not Britain's major product. The skill and versatility of weavers in Britain told in favour of semi-automatic looms, and these were retained.

Cotton exports from 1910 to 1913 contained almost twice as many linear yards of cloth as in 1870, although their value was only about 1.63 times greater. This indicates a reduction in profits, but also, since raw cotton was cheaper, in costs. Exports of cotton yarn held up well — notably to Japan which, founding its own cotton textile industry, imported less British cloth but more British yarn than formerly. British cotton thread firms, especially J.P Coats and Company, had a worldwide market and set up subsidiaries abroad rather than pay foreign import duties.

At home, however, imports of cotton cloth were over 10 per cent of the British home cotton sales by 1907. These imports were not cheap, and were often of fashionable materials, perhaps made and designed in France. In the fragmented structure of the British industry, choice of colours and patterns of cottons was often the decision of foreign-born merchants who may have been slow to react to changing tastes in Britain. Firms like Horrocks of Preston, dyeing and printing in its own works and selling direct to shops, were exceptional — and more adaptive and profitable.

As many other countries were expanding their cotton output behind high tariff barriers, exports became more difficult. The British industry had to find new export outlets away from Europe. This it did increasingly in India, the Middle East and the Far East. But India was also developing its own industry — often using second-hand machinery discarded by British manufacturers — and Japan, with its very much lower wages, was exploring Chinese markets, competing with some British and American cloth. This bred anxiety.

The switches of markets, for the time being at least, worked. Good judgements ensured an expanding market and a profitable sale of what was produced, in spite of all the subsequent criticism.

Woollen and Worsted

The trademark of 'pure new wool' on woollen cloth could not have been put on up to 50 per cent of the output of the woollen textile industry of the nineteenth and early twentieth century. It contained fibres other than wool. 'Wool' was not all 'new' — that is, wool yarn was made from the waste products of spinning and weaving authentic woollen yarn, or even from fibre unpicked and unravelled from old cloth. Yet the cloth was classed as woollen. Worsted, differently spun and woven, was even more likely to be made from mixed fibres.

The woollen industry was slower to adopt power-driven machinery than cotton. Wool fibre, not as tough as cotton, did not lend itself quite so readily to faster, jerkier machinery. But by 1871, while there was still a sizeable number of hand-loom weavers, and 13 per cent of woollen mills in England and 27 per cent in Scotland were still water-powered, steam predominated. Less than half the output of the woollen and worsted industries was exported — perhaps 30 per cent by 1907, when net output of woollen and worsted products equalled £18 million. Woollen exports were 6 per cent of total British exports in 1913. The woollen and worsted industries employed some 200,000 people at peak, though numbers fell after 1870. By the 1880s, 72 per cent of those employed in woollen manufacture worked in Yorkshire. In Scotland, where employment in the industry was at its peak in 1874, numbers declined, as also in Wales and the West country. But Scottish figures rose again after 1905, with a revival of interest in Harris tweeds, houndstooth check, and factory-made knitwear — socks, stockings, and men's pullovers and cardigans.

Fully automatic looms in the woollen industry were hardly used before 1914. The skill of weavers in setting patterns and adjusting weaves counterbalanced the advantages of automation. There was some improvement in the equipment used, but no wholesale replacement of it. In value, British woollen and worsted exports stayed more or less constant, but increased somewhat in quantity (a difficult matter to estimate, since narrower widths of cloth gave way to wider ones during the years 1870–1914). No country showed much increase in woollen exports, though the world's consumption of wool rose as successive countries established their home woollen industries. US tariffs of 1894 and 1898 dealt a severe blow to British woollen exports: replacement markets were found scattered all over the world.

Technical institutes and classes in woollen textiles were established during the 1870s: four new ones in 1878 alone. They were criticised by 'practical' men, and indeed virtually nothing was known of the chemical properties of wool. The classes were largely in weaving and design, and suffered from some rather inept teaching and poor attendance. Another reaction to hard times was for firms to merge, rather than compete. The 1898 Bradford Dyers' Association, a horizontal merger of 21 firms, was intended to keep up prices and quality, on which its reputation depended, and it had some success. The Yorkshire Woolcombers, however, formed in 1899, was a failure. Merging was not enough: organisational efficiency and realistic appraisal were essential to success.

There was increased interest in cycling, tennis, cricket, and golf. Even people who did not participate liked to look as though they might. The British woollen industry, which had a considerable degree of vertical integration, responded. Cream flannels for trousers, and lighter-weight tweeds for cycling outfits, plus-fours and Norfolk jackets were produced. The growing tide of woollen imports was reversed and, if the export trade did not expand, at least it did not fall as it might have done.

Worsted was less happy. Centred on Bradford, it had a large number of smallish firms. Of the 3,063 in 1870, only 16.8 per cent survived to 1912, though a number of new ones started, keeping numbers up to 1,638. The prime reason for the decline overall was French worsted hosiery which was all wool, flexible and fashionable, knitted on different machinery: British manufacturers could not compete.

Woollen hosiery from Belgium was also popular, and this displeased British manufacturers. On the other hand, there was satisfaction in noting that between two thirds and three quarters of British woollen yarn exports went to Germany. New developments in knitting machinery increased the vogue for factory-made knitwear, and there was new investment to produce it. Yet as in cotton, there was concentration on the use of skills rather than investment in new technology, though more response to changes in taste.

Silk

Manufacture of silk in Britain was not large. The number of people it employed fell from well over 130,000 in 1851 to 39,000 in 1901. When the price of raw silk dropped considerably, the price of silk goods did not fall commensurately for some time. It was a prestige product. One firm, Courtaulds, dominated an increasingly specialised section of the silk industry: this was black mourning crape, worn by the Court, the fashionable, the devout and the devoted, as a mark of sorrow.

Courtaulds' dominance both in Britain and in France was because of their 'crimping' process, which gave their fabric a much-admired texture other manufacturers could not imitate. The firm was a family partnership until 1891, when it became a joint stock company, but still family-controlled. Highly prosperous for many years, guarding its processes not by patents but by locked doors, imparted skills and sworn oaths, it declined in the 1890s as public taste at last veered away from such lengthy displays of overt sorrow, and as other manufacturers of crape experimented and produced similar, sometimes cheaper, materials.

But the three-generation firm did not go under. Instead, in spite of enjoying the lifestyle of wealthy country gentlemen, the family took on well-chosen new managers, new machines, electric light, and telephones, and diversified into a variety of coloured silk fabrics. Though foreign sales of crape recovered somewhat, home sales did not. In 1904, one of the new managers advised the firm to buy a patent for 'artificial silk'. This was the effective beginning of the British rayon industry; Courtaulds was not slow to seize a new process and product when profits in the old shrank.

The Iron and Steel Industries

The iron and steel industries are 'old' industries by any definition. However, in the years 1870–1914, both the products and the methods of producing them were changing. The first product of the smelting of iron ore is pig iron, which is treated in alternative ways to produce wrought iron, cast iron, and steel of varying properties. Britain in 1870 produced half the world's pig iron, largely in Northeast England, South Wales, and the Clyde region. Steel had been an expensive product, difficult to make, produced in small quantities. In 1856, Henry Bessemer, of German parentage, but born and living in England, invented a way of converting pig iron into steel. However, at a demonstration for iron producers, his iron failed to 'make' into steel. Three years later, he had established that he needed low-sulphurous coal combined with low-phosphorous iron ore. Producers were impressed, and installed Bessemer converters. The price of steel fell, demand rose, and production doubled from 1.29 million tons in 1880 to 3.58 million in 1890, and again to 7.66 million by 1913. Iron and steel output equalled £30 million in 1907.

Low-phosphorous iron ore began to run out in Britain; supplies were imported largely from Spain, but also from Sweden. Another method of making steel, the Siemens method, the open-hearth method (invented by a German living in Britain), also used low-phosphorous ore: it too had its difficulties at first, and was more widely adopted abroad than in Britain. Further developments facilitated the use of high-phosphorous ore, but the higher phosphorous ores of Britain were not much used because they were embedded in a high proportion of rock. British steel producers were criticised for not tackling this problem more energetically.

In 1880, over three quarters of the world's international trade in steel had originated in Britain. In the next 35 years, world demand for steel grew fifteenfold. It was unlikely that British output and exports could increase enough to keep up the same share of a vastly expanded world total and although British output doubled, by 1913 British exports amounted to only one quarter of world trade in steel. This decline in share was the subject of much criticism at the time and subsequently, and the industry was given symbolic significance as an indicator of Britain's industrial strength — or weakness. This in itself was foolish. US and German steel producers, with their natural resources, their newly integrated home markets, subsequent high tariffs, and immediately adjacent markets in Canada and the Rhine valley axis, inevitably grew faster than British. Output per man in the more mechanised American plants was higher than in British — *but* their wages and capital equipment were more expensive. Quality for quality, British steel sold at much the same price as German and American, unless there were divergent transport costs, or deliberate dumping. Germany and Belgium on occasion did 'dump' as their production outran demand. While accusations that German — and Belgian — steelmakers *intended* to ruin the British industry by dumping were unfounded, dumping caused difficulties for free-entry Britain.

From the 1880s, British steelmakers were cautious and conservative. Having abundant coal (though, since they owned coal-mines, they did not always trouble to cost it properly), they did not install fuel-saving devices. These might have yielded additional benefits. Their works were scattered and comparatively small-scale, reflecting their origins. A technically outstanding integrated works installed by the Steel Company of Scotland lost out in an over-supplied market for steel plates. This confirmed the caution of other steel men; their attitude may have bred trouble for the future, but not for the years 1870–1914.

A comparatively small but essential part of the British steel industry was devoted to high-grade steel. Sheffield firms used Swedish bar iron in small crucibles to produce it. From it, cutting edges of machines and tools were manufactured, and steel alloys developed for specific purposes.

Metallurgy was still in its infancy; there was no microscopic examination of steel until the 1890s. But so skilled were Sheffield men that, by breaking an ingot and looking at its crystalline structure, they could calculate the carbon content of the steel accurately to within one tenth of one per cent. R.A. Hadfield, owner of Hadfield's Steel Foundry, added 12–13 per cent manganese to steel, making it harder under pressure. He publicised this in 1888; it was eagerly adopted in the USA, where a growing machine tool industry utilised it. The scale of demand, and the readiness to use something new, were greater there. Another development, which also took root more readily in the USA, was the addition of about 12 per cent chromium to make stainless steel.

At least until the turn of the century small firms, often working for known customers, held their own in Sheffield; after that time, larger firms, sometimes absorbing the smaller ones, were more successful. However, the application of discoveries needed confidence in both the product and its market, and the application of capital, and here British firms hesitated — and lost out.

The Shipbuilding Industry

The nineteenth century saw every possible combination of wood, iron and steel in the construction of ships, and every variation in the degree of motive power provided by sail and steam respectively. The choices were partly a matter of operating costs. A reduction in the amount of coal per horsepower increased the demand for steam engines. 15–20 lbs of coal were needed per hp in the early days of steamships, 5 by 1852, 2.5 by 1859. The 1880s saw the development of the triple-expansion engine; by 1914, quadruple expansion was practicable, and coal consumption was 1.25 lb per hp.

Britain had been the major sailing shipbuilder in the early part of the nineteenth century, but was later rivalled by faster, cheaper, but less durable American ships. Steam reinstated British supremacy: it was an Englishman, Charles Parsons, who invented the steam turbine, first intended for the electricity supply industry, and the turbine gave an extra edge to British shipbuilding competitiveness by increasing the speed of British-built ships. From 1900 to 1914, 60 per cent of world shipping tonnage was produced in Britain; in 1907 the value of its net output was £21 million. Specialised production of oil tankers, refrigerated ships, and warships began: particular shipyards built for particular customers, and owners of shipping lines had directorates in shipbuilding companies and ordered from them. Innovation was cautious, but by 1889 97 per cent of plates used by shipbuilders on the Clyde were steel, not iron.

Naval orders were important to shipyards: 10 per cent profit on them was thought reasonable by all parties and in many instances this in effect subsidised the building of merchant ships. Naval influence on design, once considered stultifying, in fact was beneficial. A small naval School of Mathematics and Naval Construction at Portsmouth, set up under a Cambridge mathematics graduate, operated from 1848 to 1853. It took only eight pupils a year, but these men carried forward the modernisation of the Navy, and the shipyards which built for the Navy then used these improvements in merchant ships. Another short-lived naval school was absorbed into Greenwich Naval College in 1873, but the good work continued. Chairs in ship design under various titles were established at the universities of Glasgow, Newcastle, and Liverpool in 1883, 1890, and 1908 respectively. This sounds encouraging — but of 40 students per annum enrolling for the Glasgow course, only 2 on average completed their studies.

Other technical schools and classes were begun for apprentices and workmen by shipbuilders, but since they were available only after a day's work, they were poorly attended, and the mathematics were beyond the capacity of many students. Thus forward-looking design was combined with traditional work-practice. Further incentives to high-performance engines came from lucrative mail contracts, awarded to shipping lines which conformed to Admiralty specifications of performance and structure. Shipyards installed hydraulic cranes, and used a mixture of water, steam-power and electricity, usually very efficiently. The owners guarded against overinvestment, knowing how greatly demand fluctuated, and relied on skills rather than equipment for the high standard and comparatively low cost of the finished product. Family-run firms, albeit large ones like Sir W.G. Armstrong, Whitworth and Company, predominated. These owners and their families were personally trained as apprentices and were not research-oriented. Labour relations in the industry reinforced managers' doubts about introducing new work practices, as will be seen in Chapter 3.

The Engineering Industry

The engineering industries have in common the working of metal (usually iron and steel) by hand or machine tools. Their workforce from 1870 to 1913 was almost exclusively male. The skill of both workers and their managers, acquired principally through apprenticeship, reflected the fact that technical progress was predominantly incremental, that is, growing little by little.

A revolution in engineering came about through the development of high-accuracy machine tools including lathes, drills, milling and grinding machines. They were hand- and foot-powered at first; later their rotary action was powered by belts connected to a central steam-engine. Machine tools could be automated: once set they carried out a sequence of pre-determined operations to produce identical pieces in large numbers — that is, interchangeable parts for assembly.

The years 1890–1914 saw the introduction of turret and capstan lathes (which had to be set by craftsmen but could be operated by semi-skilled labour) and milling and grinding machines. The change from iron to steel for many products necessitated the use of new tools to make them. By 1907, net output of engineering products, old and new, was worth £50 million, in an industry second only to coal mining in value.

By 1914, the navy and the army had heavier and more accurate artillery than formerly, and machine guns and small arms which remained standard until 1945. These weapons showed very significant advances in engineering by the private firms that moved into armaments from the 1880s. Pre-eminent were Vickers, Joseph Whitworth & Son and Sir W.G. Armstrong. In the 1890s, two large general engineering armaments and shipbuilding firms, Vickers, Son & Maxim and Sir W.G. Armstrong, Whitworth were formed by merger. These led the world in design and engineering practice, reflected in the high exports of British armaments. Their new products, often involving the development and use of new alloys, had significant effects on process technology. Nevertheless, their management and costing left much to be desired.

British firms in textile machinery were innovative, and initially had an almost complete monopoly of the world market except in a few countries, noticeably the USA. Long production runs of standard machinery were possible. The industry, concentrated in Lancashire, by 1914 was dominated by six firms which each employed between 3,000 and 11,000 men. The largest was Platt Brothers of Oldham. Skilled craftsmen and traditional production techniques were management's recipe. By 1914, there were some signs that this might not be sufficient. The very durability of the product militated against an expanding home market: a Platt mule would, it seemed, last forever. Increasingly, the industry exported but other countries, especially Germany, were doing the same, and Britain was not in the forefront of new knitting machine design.

The men who managed railway workshops where locomotives and rolling stock were built were kings in their own country; they worked from trial and error and from practical knowledge rather than from scientific principles. They did not readily accept interchangeable parts, produced from the 1860s but not developed to the same extent as in the USA. British workshops, tied either to individual railway firms or to specialised export markets, did not produce standard engines in large numbers, and it is doubtful whether they wanted to. As a consequence, British engines were expensive, specialised, and of very high quality, enduring for 60 years or more. By 1914, British locomotives (and rolling stock) were losing out in third markets to less expensive, if less durable, US products. Moreover, world demand was slackening. Even the most successful rolling stock amalgamation, Metropolitan Carriage Wagon and Finance Co., found the going rough. By 1914, the industry had shrunk a little.

Boot and Shoe Manufacture

Making boots and shoes had been a handcraft, a workshop matter. From the 1850s, treadle-operated sewing machines speeded the process; and from 1875, welt-sewing machines, sewing the uppers onto the soles 54 times faster than by hand, were introduced from the USA into Britain. Their advantages were obvious. In spite of resistance by craftsmen, strikes, and worry about the cost of machinery, owners and managers installed new machines. Without them, the industry would have collapsed. Over the next decades, factories with 700 to 800 workers were not uncommon. Northampton and Norwich, each with a nucleus of craft-trained men, and useful connections with the leather trade, became the headquarters of the industry. US boot and shoe imports into Britain, negligible before 1894, rose to over a million pairs in 1905: employers had to impose new machinery to survive.

The unusual feature of the industry was that this machinery was almost all *leased* to factory owners, through the British United Shoe Machinery Company. It was developed in and owned by its US manufacturers: skilled shoemakers were in shorter supply there than in Britain, and customers were prepared to accept a standardised product. Leasing had some advantages to British manufacturers who did not have to provide a large capital outlay. The quality of some factory-made shoes in Britain — especially when their making was paid for by piece rates — left much to be desired, but visits to the USA taught shoemakers, and the multiple retail shoe shops which were opening in association with them, that better quality products were both desirable and necessary to increase sales.

In spite of severe competition, exports of boots and shoes tripled from 1875 to 1914, to be worth £4.5 million in 1914. The largest markets were in Australia, New Zealand and South Africa. In this instance, British owners and managers adopted new, imported techniques, and improved retailing of the finished product.

The 'old' manufacturing industries were expanding, on balance, with good sales and a good reputation attached to the product, but with question marks hovering over the methods of production and over the security of markets as other countries inevitably industrialised. We now turn to some new industries, beginning with the newer products of engineering.

NEW INDUSTRIES

The Engineering Industry

Three new products developed in light engineering, dependent on demand from rising incomes and new social relationships. The sewing machine came first. In Britain its production was dominated by the American Singer Company and 'American' technology. Its plant at Clydebank, opened in 1870 was, by 1900, the largest sewing machine factory in the world, producing 13,000 machines per week. The most advanced machine tools were used in the manufacture of the interchangeable parts from which the famous Singer machines were made. Interestingly, for years the sophisticated and specialised machine tools required were made by the company itself. Its main market was the modestly comfortable household, but it also produced for industry.

A product demanding similar techniques in manufacture was the typewriter. American, like the sewing machine, it took root in Britain more hesitatingly. In the 1890s various British firms began manufacturing typewriters, using licences for US patents and US technology. Most failed before 1914. The one successful British venture, the Imperial Typewriter Company (1903) also used US tools and designs. It was necessary to advertise to widen the market. But it was difficult to persuade established male clerks — or their employers — to change from pen to typewriter. This reluctance gave impetus to the acquisition of skills by girls and women, and a new white-collar career opened for them.

The third product, the bicycle, from the late 1870s became a speciality of British engineers. There was a strong home demand for cheap short-distance transport for leisure, and subsequently for work, and a well-developed and versatile engineering sector produced bicycle components. The 'Safety Bicycle', perfected in the 1880s by J.R.

Stanley, founder of the Rover Company, completely replaced the earlier 'Pennyfarthing'. In the 1890s numerous firms entered the cycle industry. The largest, which later became Raleigh, employed over 850 men and had an annual output of 30,000 bicycles. British Small Arms (BSA) in 1879 began the production of bicycles in an arms factory which had been closed for a year due to a lack of government orders.

Like the sewing machine, the bicycle found its market in the upper income bracket of the working class and in the lower middle class, as well as in better-off households. The bicycle added to a clamour for the improvement of road surfaces in Britain; it also added to British exports.

Motor Vehicles

In 1913 the output of all motor vehicles (cars and trucks) was 34,000, and the industry employed about 100,000 men (an indication of its labour intensity). Its products were for the well-to-do. 1896 was the year of the repeal of the famous Act of Parliament which prohibited steam or motor vehicles from travelling faster than 5 m.p.h. on public highways, and required these vehicles to be preceded by a pedestrian carrying a red flag. There is no evidence however that the Act had held back the development of motor vehicles. 1896 saw the establishment in Coventry of the English Daimler Company which, at first importing all the parts, became the first substantial manufacturer in Britain making stately vehicles for stately customers. Between 1896 and 1913, about 200 different makes of car were put on the market, but more than 100 disappeared. Many of these companies had previously manufactured bicycles: many owners and managers were car enthusiasts building for enthusiasts.

How successful and competitive was the British motor industry before 1914? British output was below that of the French and the Americans, and in 1913 Britain was a net importer of motor vehicles. In Britain itself, the American Ford company began the manufacture of Model T Fords in Trafford Park, Manchester and, by 1913, produced over 6,000 vehicles p.a., more than any British producer. Armstrong, Whitworth, which made some motors from 1905, had planned to manufacture cars on Ford's scale, and had the necessary expertise and financial resources, but decided against it because the estimated profit from 6,000 cars p.a. was less than that from the sale of a single river gunboat. However Vickers, Son & Maxim in 1901 acquired Wolseley, an existing manufacturer, and ran it successfully. In 1910 BSA took over Daimler.

All companies bought in components, and here there was evidence of some lack of enterprise. William Morris, previously a bicycle maker and repairer, had problems getting suppliers to provide parts to the standards and number required when he began making cars. Attempts to manufacture spark plugs foundered in the face of competition from the USA. Magnetos came from Germany. Only Ford's US factory had a conveyor belt before 1914. Motor car assembly elsewhere was essentially a handcraft and early cars were really adapted carriage designs with an engine built in. By 1914 the car was distinct from the carriage, and the motor car and the light van were finding a commercial market. Charabancs (buses) for commercial outings also appeared.

Electrical Engineering

By the 1870s electricity was widely used in telegraphs which linked all large British cities. The railway companies who initially owned telegraphs depended on them for organising traffic, and the major financial and trading centres at home and abroad were linked by overland and submarine cables. The telegraph system originally relied on chemical batteries for power; but the development and improvement of the dynamo permitted the efficient production of electricity from steam. The first electric lamps, the very bright carbon arc lights developed in the 1860s, achieved limited use in special places such as lighthouses and the Marquess of Salisbury's dinner parties.

The wider use of electricity resulted from the simultaneous perfection of the incandescent lamp in 1878/79 by the American, Thomas Edison, and the English manufacturing chemist, Joseph Swan. The Edison/Swan lamps gave a less bright light and were an immediate success. Edison and Swan were able to exploit their patents for light bulbs until 1893. Thereafter competition was intense and the British market became dominated by foreign firms and by imports, except for GEC which adopted European improvements under licence.

By the 1890s district power stations were established, serving a few thousand consumers. Uses were found for electric power in electric tramways and railways (which made the extension of the London Underground possible) and as a motive power in manufacturing. Special uses emerged in the chemical industry. Parson's steam turbine reduced the quantity of coal needed to produce a given amount of electricity from 10 lbs in the early 1890s to 4 lbs in 1914. However, the turbine was not taken up as rapidly by British electrical companies as by US and German companies and Britain, well supplied with gas, did not electrify as quickly.

By 1900 the original British electrical machinery industry had been almost entirely eclipsed by US and German firms producing for export from a larger domestic-based market, and with more trained electrical specialists. US firms established British manufacturing subsidies — Westinghouse in Trafford Park and General Electric in Rugby (as British Thomson–Houston). British companies manufactured on a smaller scale. The General Electric Company (not to be confused with General Electric of the USA), in addition to manufacturing light bulbs, produced electric motors for industrial use. But by far the largest manufacturers in Britain just before 1914 were Siemens (German) and the subsidiaries of US companies.

Cable-makers were better established, since Britain was a world leader in telegraphy, both land and submarine. The industry was dominated by two firms: British Insulated Wire, and Callendars (later merged to form BICC). Technical progress was incremental rather than radical, and this perhaps suited the British tradition of learning by doing, and helps explain British success in this product.

In 1907 total output of electrical manufacturers was worth £14.1 million. Future sales depended on future cheapening and expansion of electricity supplies.

The Chemical Industry

In 1870 the United Kingdom had the largest chemical industry in the world, based on the manufacture of alkali which used the long-established Leblanc process. The principal alkali was soda ash (sodium carbonate), used as a raw material or converted to other alkaline materials for the textile, soap, glass and paper industries. About half the 1878 output was exported.

The industry was concentrated in three regions: Merseyside, Glasgow and Tyneside. They had nearby deposits of coal, salt, and limestone, all needed in the Leblanc process; all were disfigured by its waste products. Improvements were made to the Leblanc process but, by 1914, two new processes had overtaken it.

The first was begun in 1865 in Belgium by the brothers Solvay. In 1872 two young men living in England — Ludwig Mond, a German chemist trained in Germany, and John Brunner, born in England, but the son of a Swiss scientist father — took up licences for the Solvay process in England. They met while working in the Leblanc industry on Merseyside. In 1873 they set up a private partnership, Brunner, Mond & Company. With difficulty they raised enough money locally to construct a plant at Winnington in Cheshire. They faced increasing competition, including that from the other Solvay licence holders in Britain, and in 1878 took them over. Thereafter they prospered. In

1881, they became a limited company; shares were bought by local people and by Solvay et Cie. Brunner, Mond was soon market sharing and price fixing with the now worldwide Solvay network. Solvay output was cheaper but more profitable to its makers than Leblanc.

Prices of alkali fell by 50 per cent between 1875 and 1890. All the major British Leblanc companies to survive united as the United Alkali Company (UAC) Limited, the largest chemical firm in the world. The UAC had some success, but its output declined. It was largely the sale of by-products which kept it in existence. By 1890, Brunner Mond made more soda ash than UAC and by 1907, almost five times as much. Bleaching powder (hydrochloric acid or chlorine), needed in vast quantities by the textile and paper industries, became the most profitable part of UAC's output, produced as a by-product of soda ash. New electrolytic processes also made chlorine and these, too, threatened UAC. UAC created a central research laboratory whose head was Dr Ferdinand Hurter, a Swiss, educated in Switzerland and Germany. He made major improvements to the Leblanc process, and to chlorine production, but his advice to the Board of UAC not to manufacture chlorine electrolytically turned out to be disastrously wrong. Electrolytic Kellner patents were owned by Solvay et Cie who came to an agreement with the owners of the rival Castner patents, and in 1895 the Castner–Kellner Alkali Company was established in Runcorn, producing caustic soda and bleaching powder. There were thus three competitive processes, and an expanding output.

The first important synthetic dyestuff, mauviene made from coal, was discovered by an English chemist, W.H.Perkin, in the laboratories of the Royal College of Chemistry in London. His father owned a small chemical works, and would have put his son's discovery into production, but could not find enough trained chemists in England to produce it. The works closed in 1874. Perkin went on to make a fortune from licensing his patent for this and other dyes to overseas firms. Small dyestuffs manufacturers in Britain were established and staffed often by foreigners, especially German-trained ones. Levinstein and Co., for instance, was for some years owned partly by and obtained technical information from the German chemical firms Bayer and Agfa. Britain produced 5,000 tons of dyestuffs in 1907, but consumed 23,000. Germany produced 135,000 tons, but used 20,000. This output was the basis for its new synthetic pharmaceutical and photographic materials products. Indisputably, the lack of basic and applied chemistry and research, and the lack of British firms able to recognize this, and give status and financial backing to research, was a barrier to development in this industry. Brunner Mond was an exception, but could not develop all possible products.

The modern explosives industry was based on dynamite and detonators, invented by Alfred Nobel, a Swede who later founded the Nobel Peace Prize. He set up companies in a number of countries, and in Britain established one near Glasgow, in 1873. Unusually for its day, it was a limited liability company, and run by professional managers. From 1877 it was called Nobel's Explosive Company and in the 1880s local shareholders were bought out by the international Nobel Dynamite Trust. This meant that it benefited from agreements to limit competition. New explosives were developed on the Clyde and in the Trust's other works for specialised uses, from quick-fire machine guns to earth-moving explosives, without which the Panama Canal could not have been built.

Therefore, in many products, and with considerable input from foreign-trained scientists and foreign management attitudes, the British chemical industry was, by 1914, competent and profitable, but without much development in dyestuffs and fine chemicals generally.

The Synthetic Fibres and Textiles Industry

The development of artifical silk (rayon), initially slow, gathered pace after 1880. The first commercially viable artificial silk was produced in France, by the Chardonnet process; by 1900 it was marketable: less expensive than silk, but still rather expensive.

Wood pulp emerged as the main raw material. Steeped with a chemical additive, it was further treated to produce a viscous liquid (hence the name 'viscose' applied to rayon), which in turn was extruded and set into a thread. The patent which Courtaulds bought in 1904 gave it sole British rights to make the substance in a combination of processes developed in a curious establishment at Kew, best described as an experimental workshop. Courtaulds found that much remained to be done to produce uniformly reliable, commercially viable thread. A factory was built in Coventry, a site chosen apparently because it had, after the demise of the silk ribbon trade there, a nimble (female) potential workforce, water (from a canal, by agreement) and coal.

Two chemists supervised what was in fact a chemical factory with 2,000 employees. A German process, the Müller patent, seemed to hold the key to part of the production chain; but judgement in an English court in 1907 barred Courtaulds from using it. The ban gave Courtaulds the incentive to develop by ingenuity and perseverance a somewhat similar but superior process of its own: continental competitors hankered after Müller's patent, and fell behind.

At first used primarily for braids and fringes, in a short time

Courtaulds' viscose rayon became stronger and more uniform and was woven into cloth in its Essex factories. But Courtaulds' big success was the sale of yarn. They bought out a US firm and entered into market, patent and licence agreements (which were continually broken) with European producers. Gross profit from viscose yarn, already eclipsing that from all Courtaulds' fabrics combined by 1909, was more than three times higher by 1912. The US subsidiary was even more profitable than its parent. By 1914 Courtaulds' yarn output was 3.8 million lbs; total world output was 12.25 million lbs. Here was one 'new', science-based industry which, in the hands of a single, hard-driving management, succeeded, improving its knowledge as it went, in the breathing space afforded by the Müller decision.

Conclusion

As other countries industrialised, British manufacturing industry continued to expand, though not at the rate of economies making large initial gains. Comparative advantage — even in something as elusive as French trend-setting in cotton textiles — was bound to produce national specialities. Resource endowments had a major and inevitable effect on the iron and steel industry. The remarkable British achievement was that every major manufacturing industry grew, even through the 'Great Depression'. Much criticism was misplaced.

Owners, managers and their agents found markets. Capital could be found, as Brunner and Mond showed, by the persistent. Effective management was possible, even in a third-generation family firm such as Courtaulds. Mergers, sometimes taken as a measure of 'modernity', were decided upon by managers although, as UAC showed, they were no guarantee of success. Managerial reliance on skills rather than on capital investment on the US or German pattern resulted in Britain producing quality yet price-competitive goods. All this needs to be remembered.

The undoubted deficiencies were the lack of research and of applied scientists, and the failure of too many managements to appreciate this lack. This was not for any 'decline of the industrial spirit', but was rather a continuation of the ethos of the first industrial revolution. Then, the skills of the practical man, modifying what already worked and experimenting in a hit-or-miss way without benefit of theory, had been admired. Times had changed, and newer products had different requirements. A high proportion of innovators in Britain were of foreign extraction and education because few young Britons had the necessary blend of applied scientific knowledge and financial acumen to start up profitable firms in new industries.

Older industries reaped the benefit of large-scale production, even if the efficiency of their management was questionable. It took more money to begin a sizeable factory in 1900 than in 1800: it was not always possible to start up in a small way, as once it had been, and this limited domestic competition.

Further, new opportunities beckoned as the civil service was opened to all comers by competitive examination and as banks and commercial services expanded. Such places offered clean jobs with promotion prospects to those who might have been the self-made manufacturers of the past. It was not that manufacturing was less alluring — poets, painters, and country gentlemen had never though much of it — but that more alternatives to it now appeared for the ambitious.

Further Reading

Specific Industries

Byatt, I.C.R. (1979) *The British Electrical Industry, 1875–1914*, Oxford University Press.

Jenkins, D.T. and Ponting, K.C. (1982) *The British Wool Textile Industry, 1770–1914*, Heinemann and the Pasold Research Fund.

Payne, P.L. (1979) *Colvilles and the Scottish Steel Industry*, The Clarendon Press.

Specific Firms

Barker, T.C. (1977) *The Glassmakers. Pilkington: the Rise of an International Company*, Weidenfeld and Nicolson.

Coleman, D.C. (1969) *Courtaulds: an Economic and Social History*, Vols. I and II, The Clarendon Press.

Reader, W. (1970) *Imperial Chemical Industries: A History*, Vol. I, Oxford University Press.

Questions

(1) What were the challenges to 'old' British industries and how did they respond to them?

(2) Did the British economy provide the requirements of 'new' industries?

(3) Describe and evaluate the means by which any individual firm kept or enlarged its market share for some or all of the period 1870–1914.

3

Employees, Trade Unions, and British Manufacturing Industry

Employment is a two-sided bargain. Nevertheless, it is very often a very unequal bargain. Choice about how many, and who, shall be hired — and, with the partial exception of some very recent laws, who shall be fired — does not rest with the would-be worker. In the nineteenth century and beyond, employees as individuals, unless they had some highly unusual skill, had almost no chance of altering hours, wages and conditions on which their job was offered or of making changes in it once they were engaged. Other opportunities for work in a given locality, particularly in a given line of work, might be few, and the prospect of moving daunting. Their choice was limited. It is not surprising that employees joined together in trade unions. They believed that, if all similar employees in a given firm, better still in a given region, and best of all nation-wide, banded together, with the actual or implied threat of withdrawing their labour if reason failed to persuade their employers, the balance might be redressed to some extent. In manufacturing industry, in the years 1870–1914, new trade unions were formed, and old ones increased their membership.

Employers were not hostile to all aspects of unions. Craft unions (that is, unions of skilled people) endeavoured to keep up the level of skill of their members, in part because this increased their bargaining power. Employers too had an interest in the good exercise of skills. Further, employers found it easier to bargain with an accepted or recognised trade union representative than to settle pay for each individual separately. Again, where firms of a like kind were clustered together and unions had a high percentage of possible members within firms (this is called *density* of membership), employers, seeing that if they all agreed to a certain thing each would be no worse off than his competitors, agreed to schedules and lists which established the pay for particular jobs.

In the late nineteenth century alliances, as they were called, were negotiated in a number of industries, and went further than agreements on payments. The Bedstead Alliance of 1891 between employers and workers in the making of brass bedsteads agreed that the members of the Bedstead Workmen's Association would work only for members of the employers' association and have a wages board jointly with the employers. By 1899, prices and wages in the brass bedstead trade were double their 1891 level. Furniture shops then switched to the sale of cheaper wooden bedsteads, foreign competition appeared, and some renegade employers cut their prices. This alliance, and others, could not survive a changed market.

Other employers were more antagonistic to unions: some unionists came to the view that what was wanted was not so much a change in pay and conditions as a complete restructuring of the economic framework of Britain. Yet the substance of disputes was usually immediate and real enough and centred in the workplace; and the organisation of unions, with their variety of paid and unpaid officials, was geared to the possibility of strike action, as well as to negotiation and to the distribution of benefits. Equally, employers' associations, unsuccessful as they often were in fixing markets and prices, had opposition to trade unionists' terms in mind as a feature of their organisations.

Trade Unions, Workers, and the Law

The law relating to industrial relations, to the structure and rights of trade unions and to the rights of workers has been the subject of intense controversy. Acts in restraint of trade were used against trade unions, but never against employers. The British pattern of trade union law was unique in that many rights and duties were not positively defined: rather, trade unions were granted 'immunities' from prosecution under the criminal and civil law for certain offences, and laws in general have specified what they must *not* do. There has never been a 'right' to withdraw labour in Britain; there has instead been an immunity from prosecution (though not from dismissal) for breach of contract.

The Trade Union Act, 1871, passed by Gladstone's Liberal government, gave better legal status to unions and protection to their funds, even if used for strikes. But picketing was hedged about by clauses which almost forbad it, and planning a strike was a punishable 'conspiracy'. Two more important statutes were passed in 1875 by the Conservatives under Disraeli. These extended the immunities to 'conspiracy' — that is, it was now lawful to plan strike action, as long as the actions taken were not criminal in themselves — and the scope of peaceful picketing was extended. Breach of contract became a civil offence only, with civil damages as the sole penalty.

Statute Law relating to trade unions remained the same for over thirty years, but trade unions came under increasing challenge. Judges hedged them in through a series of adverse judgements — against picketing in the 1890s, and against striking with immunity, most especially in the Taff Vale Railway case of 1901. Here the Amalgamated Society of Railway Servants was held liable for civil damages for losses incurred by the employer during a strike, and fined the then huge sum of £42,000. Resentment caused by this gave a great boost to the newly formed Labour Representation Committee of 1900, helped change its status to that of the Labour Party in 1906, and led to the passing of the Trade Disputes Act of 1906 by the new Liberal government. This restored immunities from prosecution for civil damages which the unions had enjoyed before 1901. It also consolidated previous immunities and gave further protection for peaceful picketing. The Conservatives had argued for the establishment of a positive legal code for trade unions in 1903, but this was rejected, partly because of the dislike by trade unionists of judges intruding into the affairs of trade unions.

The trade unions affiliated to the new Labour party gave it money without interference from the law, protected, they thought, by the 1871 Act. In 1909, however, the Law Lords declared that it was illegal for trade unions to give money to a political party; the Trade Union Act of 1913 made it legal again. But the money had to come from a separate 'political fund', from which trade union members could contract out, and which could only be established if the majority in a union decided by ballot that they wanted one. An incidental spin-off from the suspension of payment to Labour MPs from trade unions' funds was the introduction of payment of *all* MPs from public funds, thus removing a barrier to parliamentary candidacy and membership.

Other, more positive, changes in the law included the Conciliation Act of 1896 which allowed government-sponsored arbitration tribunals. In that year also, the Board of Trade established its Labour Department which collected national statistics for industrial disputes for the first time. The Fair Wages Resolution of 1891 instructed public authorities to purchase only from firms which observed standards of wages and conditions determined by collective bargaining. Another advance was the Factory and Workshops Act of 1901, which consolidated the previous jumble of Acts. New acts dealt with industrial accidents: the Workmen's Compensation Act of 1897 (a small beginning) put liability for industrial injury on employers.

These laws increased the rights of workpeople, though their leaders were on occasion very wary of the motives behind them, and suspicious of their effects. A strong body of union opinion held that hours, wages and conditions were best settled with employers by free collective bargaining. Trade unionists were in Parliament as Labour members to protect unions, not to do what unions could do for themselves.

The Growth of Trade Union Membership

The Trades Union Congress (TUC), initially a meeting to debate carefully prepared papers, was first held in 1868, a rather *ad hoc* affair, with 34 delegates. At the second annual Trades Union Conference (1869) about 250,000 workers were represented. By 1914, ten times as many workers, 2,682,357 out of a total of 4,145,000 trade unionists, belonged to trade unions affiliated to the TUC. Within a more assured legal framework union numbers rose. While other factors were involved, legalisation of funds and activities mainly helped this rise, and by 1914 trade unions included some 40 per cent of the male manual working population, although there was considerable variation in 'density' of trade union membership from one industry to another. In 1870 members were almost exclusively skilled tradesmen or craftsmen; by 1914 many unions had organised semiskilled and unskilled labour.

Skilled workers joined trade unions both to improve bargaining power with employers and to participate in union benefits which included payments to sick and out-of-work members. But these benefits were only available to better-paid workers who could make adequate weekly payments to union funds. For unskilled workers, as indeed for skilled workers, trade unions had shown that collective action could, at least temporarily, secure higher wages, standardisation of wages in a given locality and reductions in working hours. This was their source of strength. It was also a principal reason behind the strong anti-trade unionism of many employers, particularly after 1890. Employers were said to 'recognise' trade unions when they agreed to accept trade union leaders as negotiators on behalf of their members.

By 1914 most of the essentials of the trade union movement as it is now known were established; most major unions in manufacturing today have their origin before 1914. Contemporaries were familiar with demarcation disputes, sympathetic action, trade union immunities, the political levy, and picketing. But how much effect did trade unions have in the longer term on hours, wages, and the speed at which new technology was introduced?

Working Hours

The density of trade union membership certainly correlates quite well with reduction in the number of hours worked in a normal working week. People working for a wage in callings which had low density or no union organisation worked the longest and most sporadic hours — labourers who had no special skill which they might threaten to withdraw; shopgirls (still working over 70 hours a week in fashionable department stores in the 1890s); domestic servants; and immigrant workers in the small workshops of London's East End. Where work was scattered, and workpeople difficult to organise, hours were highest; skilled workers worked shorter hours.

Where hours were reduced by union action, this was often done piecemeal at local level. Then, what became standard in a given district, was a yardstick which could be applied in others. The 52–56 hour week of the favoured few had become the norm by 1914.

Wages

The development of collective bargaining by unions had frequently standardised wages for a given class of worker within a particular district. Conventions surrounded the relative level of wages, which did not necessarily relate to the comparative value of the work done, but reflected and perhaps reinforced social relationships. Young people earned a percentage of adult wages, often increasing with each birthday. Women earned less than men, even for similar jobs, though comparisons are difficult, since so many occupations were habitually regarded as sex-specific. These relativities stayed fairly constant between 1870 and 1914, with adult men earning 25 to 50 per cent more than women in the same occupation. Skills enhanced pay, and though there were all sorts of variations, unskilled workers in a given industry usually earned more if the higher grades of work in that industry were well-paid.

Wages differed from earnings, which could include additional elements for piece rates or extra hours, or could be less than wages if shorter hours than usual were worked in a given week or month. Wages for the same job varied from region to region, though this gap narrowed between 1870 and 1914. This makes it harder to estimate a useful national average for a given type of work. Nevertheless, some general levels were observed. Labourers, doing jobs like loading and unloading in factory yards, earned about £1 weekly in 1901. Men with some skill earned £1.25. Skilled men earned up to £3 a week, but this was comparatively rare unless they also had extra responsibility as a foreman or supervisor.

At the beginning of the nineteenth century, whole families had gone to work in cotton mills, and the father was paid a wage which he distributed as he saw fit. Gradually, wages were paid to individuals rather than to small teams. Some employers tried to introduce, sometimes very clumsily, the ideas of the US engineer F.W. Taylor; these centred on payment either by piecework, or by measured daywork. Workers were watched and timed performing their usual task. The number of such tasks which could be performed in a week was set as the norm, and earnings depended on achieving that number. Traditionally, there were all sorts of variations in wages — payments for time spent and/or goods produced — and this complicates the calculation of national averages of pay.

In addition to changes of earnings in a given occupation, there was an upward drift into occupations with higher earnings. The picture is not that of manual workers becoming white-collar workers (who, apart from the lowest-paid clerks and shop assistants, had higher pay), but of the sons and daughters of these men setting out on a career different from that of their parents. Those parents often moved into slightly better paid manual jobs. Experience entitled them to some responsibility, which then increased their pay.

Conditions of Work

The term 'conditions of work' covers two different sets of considerations. The first are environmental factors — the purity of the air breathed, the adequacy of light by which work should be done, the temperature of mills and factories, the noise level, especially in confined spaces, and the element of sheer danger in some work situations. These factors were rarely the cause of strikes. On the one hand, the worst conditions were gradually improved by legislation and by-laws. On the other, employees showed remarkable endurance in dreadful surroundings, provided other aspects of the jobs were more or less satisfactory. They had some pride in that endurance, and the skill with which they handled molten metal and other dangerous substances.

The second set of factors concerns the organisation of work — who did what, with how many helpers, the level of skill required to operate a given piece of equipment and, if that equipment should be replaced by something easier to operate, whether the same skilled man still operated it (often a trade union contention) or a less-skilled person was given the job. These factors were most contentious in the decade 1890–1900 and again in 1906–8 as employers tried to recover from the setback to trade encountered after the US financial crisis of 1906.

Engineering Unions

The engineers discussed in this and parallel sections are not theoretically trained and professionally recognised engineers, but skilled manual workers who had served an apprenticeship.

The formation of the Amalgamated Society of Engineers (ASE) in 1851 was a landmark in the history of British trade unionism. From 1870 to 1914 the Society enrolled only skilled men, mostly turners and fitters — a typical example of a craft union. It was usually politically conservative. With high weekly contributions from members, it paid sickness, unemployment, and strike benefits from funds held centrally in London.

In 1871 an engineering strike on the north-east coast, followed by further strikes in the rest of the country, established the 9-hour working day for engineers nationwide. In the late 1870s a further series of bitter strikes maintained the 54-hour week, 9-hour day, which after dropping an hour on Saturdays remained standard in engineering until 1919. At the same time the ASE pressed for uniform wages in a given district for all engineers. The Society's numbers grew considerably during the 1880s, then more slowly from 1890 to 1910.

In the 1890s the Society's members were under threat from the introduction of machine tools on which only unskilled rates of pay were given, and from the formation of employers' associations. On Clydeside, employers were using more young handymen to do repetitive machine work and speeding up output by paying more men by piece rates. At the same time, and in part in response, the Society was becoming more radical: it had changed its rules in 1892 to permit a more vigorous policy, including recruiting the machine men, and in 1896 it elected a militant general secretary. Foremen, themselves often union members, were supposed to introduce a faster pace of work, and had divided loyalties.

In 1896 the Employers' Federation of Engineering Associations (later called the Engineering Employers' Federation) was formed and the Society demanded an 8-hour day, backed by a threat of strike action. The dispute began in London; the employers organised a lockout of ASE members throughout the country. It was the first nationwide strike or lockout, and was really about much more than the 8-hour day. The employers claimed that the ASE was holding back technical progress, and demanded 'the power to manage' their own works. This is shown in the Terms of Settlement to which the Society was forced to agree in the beginning of 1898, after 30 weeks without work and 35,000 on strike in 702 firms. It had to concede to employers the right to use non-union labour, to use anyone they pleased on machinery, to introduce piecework without restriction (this covered 5 per cent of engineering in 1886, 27.5 per cent by 1906) and to have as many apprentices as they wished; and the Society lost its 8-hour-day demand. But the ASE was not destroyed: the employers agreed to establish procedures to govern conduct in future disputes through district and national negotiations. The Terms of Settlement, although somewhat modified, nominally remained in force until 1914, and provided the basis of agreement and conciliation procedures until the late 1960s.

The premium bonus system, introduced largely after 1898, meant a new emphasis on pay at the level of each plant, since it was within each plant that decisions about the time for each job were taken (and rewards given for extra speed). In 1908 engineers went on unofficial strike in the Northeast, against the advice of their London headquarters, to resist pay cuts and to demand a minimum wage that would not depend on the 'state of trade'. The strike lasted for five months and was lost: it was a bad year for trade, and the employers held firm.

Nevertheless, the outcome of the years 1870–1914 was that skilled engineers remained highly paid. Further, employers, already hesitant about introducing new machinery in uncertain times, were henceforth reluctant to press for changes, accepting that the skill of their men and an absence of prolonged disputes was a better combination than installing more modern capital equipment and different work practices with risk of disruption. Some of the terms of settlement were never put into effect.

The paradox was that many of the most highly paid men became, after the 1890s, the most radical. E.J. Hobsbawm (see reference at end of chapter) sees the 1890s as the crucial period in which the position of the highly skilled engineer was threatened by, rather than enhanced by, the ongoing industrial revolution. Hence, for the first time, this key section turned against capitalism. Almost from the start of the late nineteenth century movement engineers were prominent among the socialists. At the ASE's Delegate Meeting in 1912 more than half the delegates present appear to have been advocates of 'collectivism', to be achieved by class war. Since they also were in the forefront of claims on wages, hours and status, they retained the loyalty of their members, whether those members shared their political ideals or not.

Cotton Textile Unions

Many unions in cotton textiles were small, but effective organisation was secured through loose 'amalgamation' of local unions. The spinners, better organised than the weavers, were skilled men, but they had learned by watching and doing, and had not served an apprenticeship. Typically, in 1870, the spinners employed their own assistants, often members of their own family. They therefore had a great deal of control over their work and over their earnings, and probably had the highest level of unionisation. It possibly reached 100 per cent before 1914, and had the highest levels of contributions and benefits in the whole trade union movement. Among the other workers in spinning, unionisation increased, but never to the level achieved by the senior spinners.

In 1891 the Federation of Master Cotton Spinners' Association was formed (employers), and in 1892 it organised an effective lockout of the cotton unions to enforce a 5 per cent reduction in wages as profits were squeezed in slackened trade. The lockout lasted five months, and was resolved by the Brooklands Agreement. This stipulated a 2.9 per cent *increase* in wages but, more importantly, established procedures for central and systematic bargaining. A dispute not settled at the mill itself was to be passed on to local officers of the employers' association and trade union, and if need be to the national organisation. Furthermore, general wage changes were to take place only once a year, and were limited within a range of 5 per cent.

In cotton weaving, where many women were employed, progress towards collective bargaining was slower, but by 1892 the unions had 50 per cent membership density and secured uniform piece rates throughout Lancashire in the Uniform Plain Cloth List of piece prices. Indeed, the cotton unions in weaving and certain processes in spinning had by far the highest level of female unionism in the country: by 1900 three fifths of all female trade unionists were in the cotton unions.

The Lancashire textile workforce, through the United Textile Factory Workers' Association, effectively lobbied both Liberal and Conservative MPs. Union officials were wary of the introduction of ring-spinning, but did not share the radicalism of their engineering counterparts, perhaps because their own position was not under such a threat from new methods. Nevertheless, they changed allegiance from the Liberal to the Labour Party in 1911. They were resistant to change in their intricate wage structure, all of which they had to learn to become officials, and this reinforced existing work practices.

Shipbuilding Unions

The principal trade union in shipbuilding was the United Society of Boilermakers and Iron Shipbuilders, which grew from 30,000 in 1889 to 48,000 in 1900. Many other unions organised the specialised trades engaged in the varied tasks in ship construction. Changes in ship design provoked demarcation disputes over which trades should undertake new tasks, especially in the years 1888–1893. In 1890/1891 the owners organised a lockout to compel the unions to come to an agreement between themselves, but no effective solution was found because the ASE would not join the Federation of Engineering and Shipbuilding Trades. This was formed in 1890 and made up of thirteen unions with members in the shipbuilding industry. Nevertheless, shipbuilding had a well-developed system of collective bargaining; formal and informal agreements were reached about the number of fully-skilled men, the number of semi-skilled, and the number of handlers of materials needed for each task. Yet the scope for disagreements was obvious.

Once a ship was launched, there were likely to be lay-offs of labour until the yard was prepared for a new keel. Bad weather, too, added to the uncertainties of employment, since all construction was in the open air, and some winter conditions made work impossible. This type of job insecurity understandably reinforced the tendency for each trade union, and each subdivision in it, to claim work for its members while it could. Fortunately, the demand for British ships was high, and the industry could carry some of the inefficiency and loss of working days which the situation produced. Some merchant ships were built at a loss, with a shipyard's solvency maintained by its contracts with the Navy. Increased competition from abroad was likely to show up weaknesses in the industry and in the unions' position.

Iron and Steel Unions

Like cotton, but unlike shipbuilding and engineering, the iron and steel unions were skilled, but were not, strictly speaking, craft unions: members were not required to serve an apprenticeship. Seniority determined their work. In the early 1870s, a great number of local unions were formed. In the 1880s, the two principal groups were the Association of Steel and Iron Workers of Great Britain and the British Steel Smelters Association.

In 1869, the Board of Conciliation and Arbitration for the Manufactured Iron Trade of the North of England was established, and in the Northeast, often the starting place of disputes, conflict diminished.

Although many employers did not formally recognise trade union leaders, the workmen on the Board were often union members and officials. As in cotton-spinning, a system of internal sub-contracting prevailed. Senior workers were paid tonnage rates — in effect piece rates, which depended on the output (in tons) of the steel furnace on which they worked — and they then employed their own assistants on a day-rate basis. The price per ton was decided by the Board and varied on a sliding scale with the price of iron. In general, the Boards worked well.

During the transition from iron production to output predominantly of steel, the subcontracting system declined, direct piece rates were introduced for more employees and at the same time new trade unions were established at local level. They did not all join the two larger groups. However, their presence, even though it fragmented collective bargaining, did not lead to multiple disruptions. This was firstly because wage costs were not a high proportion of total costs and so employers were likely to yield under pressure; secondly, promotion by seniority and pay related to output meant that senior workers had a vested interest in keeping up production. Thirdly, the people who lost work at a time of slack trade were often not the men in the iron and steel works, where at least some production was necessary to keep the plant in good condition, but the scattered outworkers and firms involved in further processing. The iron and steel industry had a deserved reputation, by 1914, for relative harmony.

Conclusion

Industries' records in industrial relations differed, determined in part by their history, in part by the market circumstances encountered in each. The record then became part of their history and tradition and affected future developments. Since we are primarily concerned here with manufacturing, 'general' unions, with some members in manufacturing but their main membership in transport and other service industries, have not been discussed. It was these unions which grew rapidly from 1909 to 1914, and which were involved in the increased wave of strikes from 1911 to 1914, but they had some influence on attitudes in other industries.

The strike wave was the product of a combination of circumstances. These included the adoption by some union leaders of syndicalism, and stagnant money wages at a time of marginally rising prices, which persuaded many working men to accept the leadership of radicals. Syndicalism had a twofold ideology. One part related to its aim — the ownership and management by workers of their own industry (there were some differences among dockers about whether this was to be dock by dock, or through all docks joined together). The other part related to its means — the amalgamation of all workers within an industry into one union, strike action by that union, and a coordinated strike by all such unions simultaneously. Not even the first of these means had been put into effect, but men followed their leaders, since in the interim the leaders also agitated for higher pay. In 1912, a total strike in the London Docks, led by Mann, defied the Port of London Authority. Blackleg, that is, non-union labour, was called in to break the strike, and in any case other ports did not support it, and it collapsed.

Hostile attitudes affected employers and employees not involved in the disputes and perhaps heightened class consciousness. But there was a diversity of perception and interpretation based on the great variety of British working life. The socialists of the engineering unions had little in common with the syndicalists; societies which had explicitly socialist aims had a small membership, and disagreed with each other. Undoubtedly there was a widespread desire for something better than the existing state of things, but not, as small membership of socialist clubs and societies shows, a widespread wish for fundamental change in the capitalist nature of economic life. The huge gap in life style between owners and low-paid workers, sociologists tell us, is less likely to cause strife than much smaller perceived differences in standards of living of similar groups and those, to an extent, were closing.

On the one hand, the events of 1911–1914 left employer animosity to unions as their legacy; on the other, the years 1870–1914 saw acceptance of the legality of unions and of their funds, including political funds, of picketing and strikes, even of union incorporation into the government scheme of national insurance in 1911. Trade unions were the largest contributors to and provided most of the members of the new Labour Party. Their role in securing a reduction of hours and in at least maintaining their members' share of national income, or increasing it if one includes the whole period 1870–1914, has been questioned: certainly their members and their employers thought they had that role, and in some instances they undoubtedly did. While British industries remained profitable, increased wages were not detrimental to the economy, which benefited from domestic consumers with increased purchasing power. The question marks were for the future. They hung over that profitability, resistance to change and intricate patterns of rules exemplified in the shipbuilding industry. These matters should be borne in mind when thinking about the complaints, criticisms and controversies outlined in Chapter 2.

Further Reading

Clegg, H.A., Fox, A. and Thompson, A.F. (1969) *A History of British Trade Unions Since 1889, Vol. I, 1889–1910*, Oxford University Press.

Pelling, H. (1963) *A History of British Trade Unionism*, Penguin Books.

Wrigley, C. (ed.) (1982) *A History of British Industrial Relations, 1875–1914*, Harvester Press.

Specific Occupational Groups

Hobsbawm, Eric (1984) 'Artisan or labour aristocrat?' *Economic History Review*, 2nd series, Vol. XXXVII, No. 3, pp. 355–372.

Jefferys, James B. (1946) *The Story of the Engineers, 1800–1945*, Lawrence and Wishart.

Questions

(1) What advantages were trade unions able to secure for their members, 1879–1914?

(2) Why were industrial relations more hostile in some industries than in others? (Use examples.)

(3) In what ways, and to what extent, did laws and governments become involved with trade unions, 1870–1914?

Section B

1914–1939

… in the Coventry Ordnance Works

4

British Manufacturers and the Economic Environment

Of all advanced economies, the British in 1914 was the most committed to international trade and finance. Any major international upheaval was bound to affect it, and affect British manufacturing industry in particular. In 1914 a major war began. The pattern of Britain's manufactured output, of its international trade and finance, of the composition of its workforce, and of the nature of government intervention in the economy was changed. The war also had important psychological effects. Some changes were temporary, others long-lasting. To understand them and their impact on British manufacturing industry we need to know about expectations of the nature of the war, its actual size and scope, the terms of the peace settlements which followed it, and the principal features of the international economy until a second major war began in 1939.

Perspective: War and Peace

The Anticipation and Reality of War

There was in Britain in 1914 considerable hostility to Germany and its naval building programme, but little serious thought about a major land war (it was a hundred years since Britain had fought one) and little popular recognition of what prolonged destruction on the high seas might do to the British economy.

British defence plans were for an army to go to France for a short time, and for a naval force to defend the North Sea and Channel and destroy a German navy in a set-piece battle, protecting trading routes. Most government plans were to deal with a short war. When a Serb assassinated the heir to the Austro-Hungarian throne in late June 1914, one by one European nations were brought into the war. The German armies which planned to defeat France in three months were stopped in Northern France, and trench warfare ensued there for four years. Germany in 1916 built a fleet of submarines which sank British and allied troop and supply ships. This helped to bring in the USA on Britain's side in 1917. The conflict was called the Great War. Subsequently, after another conflict, it was referred to as World War I, in which Germany, Austria, Turkey and Bulgaria fought Belgium, France, Britain and its empire, Russia, Italy, Greece, Rumania, Portugal and the USA. It was a vast conflict.

The greatest damage on land was in Belgium and northern France. The civilian population of Russia suffered immense hardship from food and other shortages. In November 1918 Germany surrendered. Fighting continued in Poland and Russia until 1920, and until 1921 in the Balkans. Nevertheless, World War I is usually described as lasting from 1914 until 1918.

Probably the dead numbered 8.5 million military and 5.5 million civilians. Perhaps a further 20 million people died in 1919 in an influenza epidemic, lethal among populations weakened by lack of food. Three quarters of a million people from the United Kingdom were killed in the war, and over 1.5 million wounded, many suffering permanently. Supplies were disrupted everywhere. The psychological shock of such a war can be appreciated.

The Nature of the Peace Settlements and the Peace

By the peace treaties Germany had to return Alsace and Lorraine to France; its Silesian coalfields went to Poland. Thus three quarters of its known iron ore deposits and one third of its coal were lost. At the same time, reparations of $33 billion were to be paid by Germany to the Allies, a punishment which bore no relation to Germany's ability to pay. These payments were subsequently reduced and offset, until 1928, by loans, chiefly from the United States, but their collection and disbursement distorted other aspects of the international economy. The USA did not want reparations, but did want repayment of loans from her allies, and Britain, much in debt to the USA, tried to use the one to pay the other. Politicians' attempts to renegotiate terms added to uncertainties.

The separate new states of Hungary, Czechoslovakia, Yugoslavia, Poland, Finland, Latvia, Lithuania and Estonia emerged from the peace treaties. The government of each aimed at economic self-sufficiency, bolstering revenue by import duties. The League of Nations was established to replace war by negotiation and rational judgement as part of the post-war settlement, but had no real power.

The International Economy in Peace

There was a surge of demand postwar, paid for in part by US loans to war-ravaged Europe. Demand collapsed 1920/1 partly because some loans ended, partly because some war losses had been made good, and partly because of the drive for national self-sufficiency. 1921–24 were uneasy years. Germany was in the grip of runaway inflation and so, to a lesser extent, were other European countries. By 1925, currencies seemed more stable and the volume of traded goods was just below its 1913 level. This was viewed as part of the recovery from the war and its horrors. US production and exports of manufactured goods grew apace. The towns and cities of the USA were prosperous.

In 1928, shares in US companies were sold at prices which could not be justified in terms of the expected earnings of those companies. US loans to Germany were withdrawn to join a kind of frantic pass-the-parcel. Share prices collapsed in 1929, the music stopped, and those left holding the shares could sell them only at a fraction of what they had paid for them. Banks and companies went bankrupt, credit became unobtainable, and warehouses were full of unsold goods. The effect spread internationally as the USA imported less, and called in foreign loans to bolster domestic US needs, and international trade slumped badly. The Depression began.

Further, war-time demands for food had seemed insatiable, so farmers everywhere had increased output. Postwar they produced more and more to keep up income as food prices fell — and so their prices and incomes fell still further. Therefore they could not afford to purchase manufactured items in the quantity hoped for. A credit crisis and a run on banks in Austria and Germany added to mounting unemployment and low effective demand. The sum of all these factors was that the developed and developing (but not the subsistence) world reached a peak of unemployment (over 30 per cent in the USA) and a minimum of demand in 1932–3. Germany for a time had 6 million unemployed. World trade fell to three quarters of its 1929 volume and just over half its 1929 value by 1932. It recovered a little, but what subsequently increased investment, output, and trade was the threat of what became World War II, which actually began in September 1939.

Contemporaries were so inclined to blame the difficulties of the 1920s and 1930s on World War I that economic historians have been wary of attributing too much to it. Yet the war had an immediate effect on supply and demand and this set new international patterns: further, in Britain it diverted efforts to understand and grapple with emerging problems which returned with the peace.

Complaints, Criticisms, and Controversies

Much controversy surrounds three topics. The first is the return to the gold standard in 1925, with sterling valued at its old prewar parity to the dollar. Was sterling overvalued and was this the cause, or at least an important reason, why the old staple industries lost some prewar export markets?

The second is the role of the newer industries and their relationship to old industries. Were new industries the key to the economic upturn of the years after 1932? Would they have grown faster if Britain had not been 'over-committed' to old industries? In order to answer 'yes' to the second question it would have to be shown not only that old industries were less profitable, but that they retained resources of capital, labour, and skills which the new industries could have used more profitably. Or was the house-building of the 1920s and 1930s the real stimulus to employment and growth?

The third is the failure of successive governments to reduce unemployment. Could more have been done? Was action halted simply by the 'Treasury view' that government expenditure would crowd out private capital investment? Did some government policies in fact increase unemployment, knowingly or unknowingly? As in Section A, some comments and judgements on these matters will be made in this and the next chapters.

Features and Factors

British manufacturing industry, after the ups and downs of 1914–23, increased output on average by 3.2 per cent a year from 1924 to 1937. This was remarkable: considerably higher than the 2 per cent per annum of 1870–1913. It was not due solely to increased labour input, though that grew at 1.4 per cent per annum from 1924 to 1937, nor to capital input, which was lower than in 1870–1914, but to an improvement in total factor productivity. Perhaps management was becoming a little more efficient; perhaps the high levels of unemployment, particularly in 1930–2, meant that employers were able to be selective in whom they hired and fired. There was also an increase in power-driven machinery. A further possible reason was that because it was easier to close down temporarily the assembly line in a new industry rather than a whole mine or steelworks, newer industries shut down at short notice, minimising overheads. Also, at a time of weak demand, less productive, less profitable plant might be expected to close for good, thus raising the average. Lastly, in the old industries, some newer equipment lowered labour input. Real incomes for those in work grew; the bitterest complaints were that no government found a way to reduce the waste of energy and spirit widespread among the unemployed. At the same time, new goods — radios, gramophones, cinema films — came within the purchasing power of those in work. In 1920 and in 1937, the percentage of GDP supplied by manufacturing was just over 30 per cent; much of the service sector was dependent on its demands.

Size of Firms

During World War I, some firms swelled in size as the Government built munitions factories and gave them to large companies to manage. Postwar, most were sold off to these companies and firm size grew. At the other end of the size scale, many small businesses were begun after 1918, often by returned servicemen using their pay-off money. Many were shortlived and not registered as companies: only when they became too big for the garden shed or garage did they enter the records. The largest manufacturing employers in Britain in 1935 are listed in Table 4.1. The numbers are higher than in 1907 and reflect a general tendency to larger firm size; growth in numbers employed was greatest in middling to large firms.

Table 4.1 *Largest Manufacturing Employers, 1935*

Employers	*Number of employees*
Unilever	70,000 (manufacturing and retailing; 30,000 manufacturing only)
ICI	50,000
Cooperative Wholesale Society	46,107 (works only)
London Midland & Scottish Railway	41,301 (workshops only)
Vickers	38,532
London North Eastern Railway	36,789 (workshop only)

Source: Johnman, Lewis (1986) 'The largest manufacturing companies of 1935', *Business History*, Vol. XXVIII, No. 2, April.

Mergers and Concentration

There is no doubt that the 1920s saw a marked increase in dominance by big manufacturing firms. The 1920s witnessed a spate of mergers: not like the prewar agglomerations of many firms, but steady accumulation by big firms of medium-sized ones. Britain was imitating the merger movement in the USA, unhindered by law. Big firms were also growing internally (although often their growth rate slowed down after a merger, as though it took time to digest what had just been swallowed). The figures for manufacturing industry are set out in Table 4.2.

Table 4.2 *Contribution of Mergers to Increases in Concentration, 1919–30*

	Jan. 1919	*Dec. 1930*	*Changes due to merging*
*Number of firms**	1,263	584	99%
Share of 5 largest firms	12.6%	27.6%	58%
Share of 50 largest firms	43.4%	63.7%	68%
Share of 100 largest firms	56.4%	77.7%	77%

Source: Hannah, L. (1976) *The Rise of the Corporate Economy*, p. 112, Methuen.
Note: *These figures refer to every manufacturing firm quoted on the London Stock Exchange, major unquoted firms, and every traceable firm which became a subsidiary of an acquiring firm already noted.

'Rationalising' was the favourite word, with its implication of financial and technical improvement: doing away with inefficient plant and consolidating production in the efficient. Many new entities set about reorganising with more vigour than prewar merged firms. Mergers reduced competition in British industry, and competition was also reduced by import tariffs and because businessmen formed associations to control markets and prices. Where most such associations had been ineffectual (except against trade unions) or of short duration before 1914, the 1920s and 1930s saw some firmly-held price rings. This could be presented as a way of not driving firms out of business, thereby preventing firm closure and job loss, but it also enhanced the survival chances of inefficient firms and inefficient practices.

Location of Industry

When the slump in the staple major export industries developed and deepened in the 1920s and 1930s, demand for *all* goods and services in the areas where they were situated fell because the income of the unemployed, and therefore their local expenditure, was well below previous levels. There was then little incentive for firms looking for expanding home markets to establish themselves in the north of England or Wales or Scotland. Even the car industry, which used sheet steel sent from Scotland and Wales, grew up in the Midlands and the outskirts of London where there were skilled workers and a nearby market — or an energetic owner-manager. Some people moved to where the new industries were: miners on a hunger march from South Wales to Westminster stopped overnight in Oxford, saw Morris's advertisement for car workers, and over 30 applied and stayed. They sent for brothers, nephews and sons, as further jobs came on offer. In 1928 the Government set up Industrial Transfer Boards, initially to train unemployed miners who wished to acquire new skills and then move to where work could be found. Some 200,000 people enrolled, but only a quarter moved permanently to new jobs elsewhere. Lack of cheap housing in more prosperous districts limited moves. The schemes ended in the 1930s, when even comparatively prosperous areas had unemployed skilled workpeople. Nevertheless, the Boards did make people aware of possibilities outside their own communities.

There were two types of endeavour to take employment to people. The first was a development of Manchester's Trafford Park concept. Factories equipped with basic services were built in advance of the demand for them, using local authority and private money. They were then leased or rented to incoming manufacturing firms. The second endeavour used central government funds under the Special Areas Act of 1934 to build factories in the four areas of highest unemployment — the Clyde Valley, the Cardiff area, and limited sections of Northeast and Northwest England. Loans, and then grants and tax concessions, were made available to firms; derelict sites were cleared and communications improved. By 1939, some £2 million had been lent, and £50,000 foregone in rates and taxes. This might seem small, yet in 1938 17 per cent of all new factories were established in the Areas, compared with 2 or 3 per cent in 1932–6. £17 million had been spent. The multiplier effect (first identified by the economist Kahn) of the expenditure of wages in these Areas must have been beneficial to the demand for goods and services locally and the new factories themselves needed local services. Yet what had been the prosperous and self-confident parts of Britain were still depressed.

London's financial and government institutions made some of its population comparatively prosperous; service industries supplied their needs. New factories found both a workforce and a market in London. Its unemployment, almost double the national average in 1913, was just over half that average in 1937 as its manufacturing flourished.

Population, the Manufacturing Workforce, Employment, and Education

War loss did not markedly affect British population growth, because for over four years emigration was virtually halted, and deaths in the armed forces were deaths of men of the age most likely to emigrate. Population figures grew thus:

1911	40.83 million
1921	42.77 million
1931	44.80 million

The rate of growth continued to slow down; the birth rate declined. In the 1920s and 1930s, families of fathers who were manual workers were smaller than previously, and there was a marked national drop in family size. Infant, child and adult mortality fell. Better nutrition meant better resistance to disease. Newer suburbs, at a distance from air-polluted town centres, and increased health awareness encouraged by new infant health clinics helped, and so did health inspection in schools. The 1911 and 1921 National Insurance Acts also provided access to a doctor.

The number of children fell; the number of older people living on pensions increased. Emigration resumed in the early 1920s, but on a smaller scale than before 1914; in the 1930s there was net immigration. The smaller family eventually sent fewer young adults into the world of work, but this was compensated for by the number of women, primarily unmarried, but also a growing number of married women, who sought and found work, many having been persuaded into it during the war to replace men. So, in spite of the slow-down in population growth, the overall numbers wanting work grew fairly steadily. Unemployment was at its peak in 1932 at 22.1 per cent of the insured population and perhaps 16 per cent of the whole. It was well over 60 per cent in some localities (e.g. Merthyr Tydfil). The hard core of the unemployed were mature men.

Technical education became more widely available in the 1920s, though its quality varied; a little more science was taught in schools and universities, and more courses on business management were begun. There were six by 1939: half the students in the one at the London School of Economics were foreign.

The war showed the need for basic and applied research, and the Department of Science and Industrial Research was established in 1916 to foster it (see p. 107). For a manufacturing nation, there was still only a small research sector in industry, lower prestige for applied science and engineering than for pure science and, by US or German standards, low numbers receiving scientific, technical, or management education.

Capital

While World War I lasted, capital investment was largely in industries which met war or minimal civilian needs. The Government appealed successfully for savings for its war loans but also had to raise taxes and expand credit. There was a surge of private capital investment in manufacturing from 1919 to 1921, much of it misdirected; then from 1924 to 1937 the growth of fixed capital averaged only 0.7 per cent a year; if the sum invested per man-hour worked is taken, there was an actual decline. There was little new investment in the troubled older industries after 1921 and many of the newer, assembly-type industries were labour- rather than capital-intensive. Factories no longer built their own power supply, so this decreased capital outlay. The tendency to save fell from 1920 to 1939; while wage earners were earning more than before 1914, income from property, dividends and self-employment fell, and taxation increased. Therefore, some groups of people who had been investors had less to invest; as profits fell, firms had lower profits to plough back. These factors of demand and supply contributed to low capital investment, but are not the whole story.

The Macmillan Committee of 1930 thought that large firms could raise funds through the *stock exchange* at reasonable cost, if their prospects were sound. However, smaller companies needing less than £100,000 were not so well served, because the fees and expenses of floating shares were disproportionately high for them. Did *banks* bridge this 'Macmillan gap' between what could be raised personally and what a big firm could raise publicly? As the Depression deepened, banks renewed loans to large customers in the old industries, and even waived interest payments. Did this mean that banks then lent less to smaller and newer firms than their prospects justified? Banks replied that they did not turn away sound customers who could afford 5 per cent interest (at a time of falling prices) and that such a rate of interest winnowed out rash borrowers. In any case, the banks needed an income to maintain reserves, meet expenses, and provide some profit. *Profits* were still the main source of capital for firms, though profits declined in the Depression. The impression is that deficiencies of design or market assessment were a more important cause of small firm failure than lack of capital. Smaller firms with some potential but also some shortcomings were often bought by larger firms, who had then to alter the organisation or the product to make them profitable. It was not necessarily lack of capital which made the small firm vulnerable.

Export of Capital

During the war, the Government forbad export of capital and caused some overseas assets to be sold. Capital began to go abroad again in the mid-1920s (though even then the Government tried to restrain it), offset to a degree by incoming capital. There is as yet no exact agreement on the figures for the 1920s. But it *is* agreed that in the 1930s there was not only an absolute decline in capital exports, but Britain imported more capital than it exported. The balance had been kept in the late 1920s by higher (6½ per cent) British interest rates. This encouraged the entry of 'hot' money, not seeking a permanent investment, which could be — and was — pulled out again rapidly. Its withdrawal in 1931 helped precipitate a *financial* crisis, and for a time no new overseas shares were allowed on the London Stock Exchange. The decline in income from exported capital meant increased pressure on the balance of payments. The decline in capital exports adds to the impression that low levels of investment in British manufacturing were not due to exports of British capital.

The Domestic Market

When war began in 1914, domestic demand for goods and services related to war use rose, and suppliers increased their prices. Eventually, stung into action by a shortage of shells for guns, government controls limited price rises, dropped production of non-essential items, and increased output of essential ones. Everything which was made was sold; demand was high. Pay in the armed forces was low, but at least it was steady; casual employment and unemployment virtually disappeared by 1917 and so there were new spenders keeping up domestic demand. Wages and prices roughly doubled from 1914 to 1918, and earnings rose more than wages, because earnings included payments for over-time working in many occupations.

Government orders were scaled down as soon as war ended: slack demand was feared. Instead, there was a boom. The 33⅓ per cent 'McKenna' duties, named after the Chancellor of the Exchequer, introduced in 1915 to curb specified luxury imports and thereby save scarce shipping space, were continued and this benefited domestic sales of British makes of some items. Tariffs were added in 1921 to protect 'strategic' industries whose weakness had been revealed by the war. In 1919 and 1920 the old staple exports also flourished. Wages in these industries leaped ahead, temporarily, and were spent on British goods.

When the slump in exported goods came in 1921, domestic demand for manufactured goods declined as unemployment rose (see Ch. 6,

Table 6.1). After the early 1920s, money wages held steady as the price of imported foodstuffs fell. This fall in turn kept down prices of British-grown food, and employed people had more to spend on other things.

The average family, with fewer children to feed and clothe, had more money to spend on housing. Between 1919 and 1939, an unprecedented 1.4 million houses were built for rent by local authorities, and government help was given to build 500,000 more. A further 1.5 million were built privately, largely for owner-occupiers, making 3.8 million new houses. Prices and the interest payments on them as a percentage of occupiers' incomes were *lower* than prewar, though the poorest section of the nation could not afford such a purchase or the rent of new council houses. The numbers of heads of families (and a higher proportion than hitherto got married and set up their own households) in the income group £175–£250 a year increased. Better paid manual workers, and an increasing number of white-collar workers, swelled the throng of new home occupiers. Their smaller families wanted and could pay for electric lights and electric irons, chairs and tables, carpets (at least for the sitting room) and linoleum (for everywhere else). The houses themselves needed taps and sinks, baths, and the new metal-framed windows. Housebuilding mobilised capital, public and private, and stimulated demand, but it could not by itself *create* capital.

The domestic market was given extra protection by an across-the-board 10 per cent import duty in late 1931, levied on all except Empire or specially-exempted goods, to stem the outflow of sterling and to raise revenue. It was expanded by the Ottawa Agreements of 1932 which gave tariff preference to Empire/Commonwealth goods. Domestic demand fell from 1930 to 1933 but as some of the staple export industries revived, so it too recovered. From 1937 to 1939, because the government began to place war-related orders, demand recovered further. In 1938, all but 7 per cent of the market for manufactures in Britain was supplied by British firms, a higher percentage than prewar, though by 1939, particularly in textiles, Japan and other countries had begun to make some inroads.

Export Markets

Scarce shipping space and a government ban on trading with enemy countries, as well as enemy sinkings, of course limited British exports between 1914 and 1918. In 1918, exports were only 40 per cent of the 1915 volume. Exports then boomed but the boom encouraged some rash expansion. Exports fell away again, rallied to 1929, fell more severely in the Depression, and improved to 1939 with a hiccup in 1937/8. Both volume (which was the important thing for British employment at that time) and value were affected. By 1937, British exports were still only 77 per cent of their 1929 volume, and 67 per cent of their 1913 volume. They were 71 per cent of their 1929 value in 1937, and still below their 1913 value. World manufacturing production was by 1937 10 per cent higher than in 1929, but international trade was lower than in 1929 because more countries were producing goods for themselves. The figures for the percentage of international trade in manufactured goods for the United Kingdom and its major competitors changed as shown in Table 4.3. Britain's loss looks to be the USA's gain to 1929, but the USA was turning out a different mix of goods from Britain, though its motor vehicle exports were rivals in some markets. Faster Japanese growth, however, was in cheap textiles which undercut British textile prices, and in engineering products which undercut German sales. German exports dwindled in the 1920s, in part because of its loss of industrial districts in the peace treaties. Some German revival was due to barter agreements made in the 1930s with Balkan countries. Some, after 1936, was due to Hitler's rearmament programme, and some to carefully manipulated currency values.

Table 4.3 *International Trade, 1913–1937: Percentage of Origin by Country*

Year	*UK*	*US*	*Germany*	*Italy*	*France*	*Japan*	*Sweden*
1913	31.8	13.8	19.9	3.5	12.8	2.5	1.5
1929	23.8	21.7	15.5	3.9	11.6	4.1	1.8
1937	22.3	20.5	16.5	3.7	6.2	7.4	2.8

Source: Matthews, R.C.O., Feinstein, C.H. and Odling-Smee, J.C. (1982) *British Economic Growth, 1856–1973*, p. 435, The Clarendon Press.

Under the Ottawa Agreements, Empire/Commonwealth (largely food) imports into Britain grew faster than British exports to those countries, but at least the Agreements helped create a more stable atmosphere after the accumulated shocks of 1914–32. Tariff barriers elsewhere for manufactured goods were a major reason for disappointing export figures. Prices of imported primary products began to rise in 1934; their producers had a better income and therefore revived their demand for British manufactured goods.

Balance of Payments

Deficits on current account, 1914–18, were not large compared with Britain's resources: more worrying, in the longer run, was the trend of figures in the 1920s. Only the low price of imports prevented a gap from appearing: income from abroad fell as a proportion of GNP. This was a continuation of the pre-1914 trend, but there is no doubt that war and subsequent upheavals reinforced it. In 1926 there was a balance of payments deficit, due mainly to a decline in coal exports. In the 1930s there *were* small deficits in most years, balanced only by an outflow of capital. Had there been no war in 1939, action would have been inescapable to correct this. Payments to the USA ended in the 1930s when reparations from Germany ended. A possible drain was curbed by import tariffs.

The Role of Sterling

War disrupted international payments, leaving sterling debts unpaid in London; and the Government borrowed heavily from the USA. The Government, afraid that uncertainties would lead to massive withdrawals of gold from Britain, issued banks with £1 and 10/– (50p) Treasury notes, which promised to pay gold to the notebearers. They reached £354 million in 1920, and increased the money supply. Much more important, the Government resorted to short-term loans, the basis of credit expansion by the banks. Money supply and prices unrelated to stocks of gold rose.

Other countries also experienced inflation (for that is what the general increase in prices was), but at various rates. This caused changes in the relation of currencies to each other. Accelerating these changes during the war was the virtual ending of shipments of gold because insurance companies refused to insure gold on the increasingly dangerous seas. Britain was effectively off the gold standard, though this was not officially announced until March 1919.

The great 1919–21 fluctuations in trade, in non-sterling currency values, and in British unemployment (it rose from 3 per cent to a temporary 20 per cent in a matter of months before falling to about 10 per cent) bred uncertainty. Gold was a symbol of what looked like the enviable stability of prewar trading relationships. Politicians, businessmen, and financiers thought the pound and the dollar should be restored to their prewar parity of £1=$4.86, both backed by gold. US overseas trade and lending was expanding: the dollar was in demand. It became a matter of prestige, in part, to boost the value of sterling to its prewar dollar value.

To do this, reserves of gold were built up in Britain and funds encouraged to flow in. The Treasury did not like the higher rates of interest which the Bank of England used to attract funds, yet by the standards of the 1970s and 1980s, the money rate of interest for the rest of the 1920s was not high (from 5.25 to just under 7 per cent).

Neverthless, since prices were falling it was quite high in real terms. Overseas borrowers of sterling were asked to borrow elsewhere.

For a blend of motives Britain returned to gold in April 1924 with £1=$4.86. British prices remained higher than US and some other prices; British wages were lower than US but higher than European. Where wages were a high proportion of the final cost of British exports (in coal mining above all) employers tried to make employees accept lower wages and longer hours to reduce costs. This had been foreseen by the Government. The coal miners' resistance led to the 9-day General Strike of 1926 and wage cuts, but other wages were steady. Sterling's return to gold was followed by an increase in international trade and did perhaps help it; yet though Britain's international trade also increased, it did not do so proportionately, and the slight improvement in the staple industries collapsed in 1929 with the general slump. The possible effects of a lower value of sterling will be considered in Chapter 5.

Sterling's adherence to gold and £1=$4.86 was short-lived. In September 1931, under a variety of strains, it ended and sterling devalued. There was no immediate or dramatic effect on British exports; when they improved, many causes were at work. The dollar was devalued somewhat in 1933, the franc in 1936. Sterling devaluation allowed interest rates to come down. The main beneficiaries were house builders and house buyers. For the rest of the 1930s, sterling remained a 'managed' currency; trading agreements got round some currency problems. London financiers retained a considerable proportion of their business, but the idea that the gold standard was a recipe for stability, and the opposite idea that a low rate of exchange would automatically improve exports, especially staple exports, each received a blow; nor did the lower interest rates lead to great investment in manufacturing.

Government and the Economy

During the 1914–18 war the Defence of the Realm Act (DORA) gave the government sweeping powers to do things by regulation. After the initial free-for-all, it used these and other powers to requisition goods and buildings, to allocate shipping and supplies, to control prices and to order, purchase, and distribute goods and materials. By 1918, some foodstuffs were rationed by local committees and government bought 90 per cent of all imports, marketed 80 per cent of all food, and controlled all exports. From 1916, there was compulsion to enter the armed forces (conscription) and there were curbs on civilians leaving a given occupation (see Chapter 6). However, there was no compulsion to take up a given civilian job. Government intervention affected hours, conditions, and wages. New welfare schemes, such as crêches for the children of working mothers, had government help.

When the war ended, controls were dismantled more rapidly than intended, because the economy seemed able to take care of itself. The millions in the armed forces, provided with extra pay and an 'out-of-work donation', put themselves into employment within two or three months. After 1920, however, unemployment grew, and so did unemployment benefit paid through the National Insurance scheme, extended in 1916 and 1921. All except agricultural and domestic workers, established civil servants, railway officers, the police, and non-manual workers earning over £250 per annum were included in it and the first two categories were brought in in 1936 and 1938 respectively. Government contributions to the scheme increased, in keeping with war-time promises. Benefits were roughly double their real prewar value. Insurance payments were costed in expectation of a limited number of weeks of unemployment, and not more than 10 or 12 per cent unemployment at that.

Unemployment in the 1920s and 1930s in insured industries was much higher, at a peak of 22.1 per cent in 1932. Many workers used up all the benefit to which they had contributed; the National Insurance Fund was exhausted. A variety of government schemes then supplemented payments, or provided a very basic income if all else failed. The Poor Law disappeared in name in March 1930, and with it went any local government financial responsibility for poor people, but the new Public Assistance Boards dispensing central government money used the same officers as the old system had done. Since such money, the 'dole', was dependent on a means test, and individual officers could be demanding and insensitive in their questions, means-testing became a symbol of degradation to many hard-pressed men and women. The *political* crisis of September 1931 was caused by the Labour Prime Minister's agreement that to balance the budget, unemployment payments should be cut by 10 per cent. Financiers and the Prime Minister had panicked. Most of his cabinet resigned rather than assent. Contributions to the National Insurance scheme were increased in later years, and the cuts restored.

Governments between 1921 and 1930 tried to do more than provide training or money. To create jobs they provided local authorities with funds for roadworks and other schemes additional to normal expenditure. £69 million was spent, but the schemes took time to organise, and it was not clear that all were really additional. This added to the Treasury's dislike of increased government expenditure, especially if paid for by borrowing, and in any case the numbers employed seemed disproportionately low for the sums expended. In 1928 Lloyd George advocated a great increase in public works, simultaneous and immediate, to improve the country's infrastructure — an extended telephone system, for instance, which would be of great value to industry. His pamphlet, the 'yellow book', did not win him the next election, and so his ideas were not put into practice, but it did win him the approval of J.M. Keynes, the distinguished economist. Keynes, by 1935, was in no doubt that if such schemes could not be financed from taxation, they could be paid for by government borrowing. If necessary, a budget deficit could be incurred, to be repaid in a better year. If the deficit caused a percentage or two of inflation because money increased faster than goods, that was a small price to pay for stimulating demand. After initial help (pump-priming), it could be self-moving.

Politicians were sceptical. Labour men were wary of an alternative to their policy of 'socialisation' though it was not clear how that would affect employment. Conservatives agreed that a budget deficit was part of the cause of the 1931 crisis. They disliked borrowing. Repaying interest on the National Debt had taken 40 per cent of the government's total expenditure in the early 1920s because the debt had increased fivefold during the war, and that was not a state of affairs to which they wanted to return. Income tax, increased from 1/2d (6p) in the £ to 6/– (30p) during the war, stood at 5/– (25p) for most of the 1930s, a historic peak for peacetime, and an increase was thought to be politically impossible. Management of the economy as a whole in any case was not part of the prevailing philosophy outside the Labour party, and that was out of office after 1931. Keynes' idea was not tried.

Instead, in addition to tariffs, successive governments tried to help individual industries in a piecemeal fashion. But it was the rearmament programme which provided an economically successful but humanly sad answer to unemployment in the staple industries.

Conclusion

The 1914–18 war changed the environment in which British manufacturing industry operated yet made few immediate alterations to manufacturers' views of what constituted a 'normal' peacetime economy. There was a sense in which the British economy adjusted better to war than to the peace which followed. Individual industries had new challenges. The next chapter looks at their problems and then returns to the comments and criticisms outlined above.

Further Reading

Aldcroft, D.H. (1983) *The British Economy Between the Wars*, Philip Allan.
Alford, B.W.E. (1972) *Depression and Recovery? British Economic Growth, 1918–39*, Macmillan.
Foreman-Peck, J. (1983) *A History of World Economy: International Economic Relations Since 1850*, Wheatsheaf Books.
Peden, G.C. (1985) *British Economic and Social Policy: Lloyd George to Margaret Thatcher*, Philip Allan.
Pollard, S. (1967) *The Development of the British Economy, 1914–50*, Edward Arnold.

Advanced Reading

Middleton, R. (1981) 'The constant employment budget balance and British budgetary policy, 1929–39', *Economic History Review*, 2nd Series, Vol. XXXIV, No.2, pp.266–286.

Questions

(1) What impact did war have on the British *or* international economy, 1914–1918?
(2) Could successive governments have done more to help the British economy, 1919–1933 *or* 1929–39?
(3) How far did international developments beyond British control limit the demand for British manufactured goods in the interwar years?

5

Old Industries, New Industries, and Managerial Choices

The Ministry of Munitions intervened in most aspects of manufacturing after the shell shortage of 1915. When Lloyd George moved from that Ministry to become Prime Minister, in December 1916, intervention increased. However, civil servants did not know much about manufacturing, and businessmen brought into government departments seemed impetuous and unsystematic to civil servants — and to other businessmen. The government often found it best to deal with representative managers from individual industries, co-ordinated with government through a committee. One of the most detailed pieces of intervention entailing government negotiation, not only with employers and managers but also with trade unions, was the 'dilution' agreement. In the 'Treasury Agreements' of 1915 trade unions agreed to permit unskilled and semi-skilled people (called dilutees: usually women, but also handymen and young boys) to do work normally reserved for those who had served an apprenticeship, or at least spent some years working under supervision in factory or workshop. The price extracted by trade unionists from the government for this agreement was a tax (eventually reaching 80 per cent) on excess profits of munition manufacturers.

Managers, postwar, were free of government intervention. However, from 1929, the plight of the old manufacturing industries, worsening in the slump, seemed likely to attract intervention from Britain's second Labour Government (1929–31). Afraid that this, perhaps through subsidies, would upset what was presumed to be the self-correcting mechanism of the market, the Bank of England organised the major banks in 1930 to put capital into a new Bankers' Industrial Development Company (BIDC), which helped to fund reorganisation schemes in old industries. Although it did not use public money, it was regarded in some sense as official. The Special Areas Act of 1934 has been noted. The Special Areas Reconstruction Association Limited (SARA) (1935) assisted small firms or small subsidiaries of larger ones to settle in the Special Areas. The Nuffield Trust and the Treasury Fund had similar aims. Their combined capital — which was lent, not given — was about £5 million. This was a further (not-fully-acknowledged) departure in principle from non-intervention or economic *laissez-faire*, and manufacturers accepted it.

Britain moved away decisively from the free trade that had existed for over sixty years. To advise on further tariffs for individual industries, a permanent body, the Import Duties Advisory Committee (IDAC) was set up in 1932 and both government and managers were on it. Its advice was usually taken.

The fluctuations which most managers experienced in output and sales 1914–1939 made it almost impossible to find a satisfactory 'normal' year. Moreover, every manufacturing industry performed differently. Manufacturers and employees of the 1920s and 1930s were very aware of money price change; there was still an idea that the prices, wages, and trade conditions of 1913 were the 'real' world.

The Federation of British Industry (FBI), founded in 1916 as an association of employers and organisations, had such divided opinions among its members that it found it hard to have a firm view on anything, particularly tariffs. By the 1930s it was a little more coherent, with emphasis on government support rather than government initiative, and it also tried to help itself. It sent trade negotiators, reasonably successfully, to Japan and the USA in 1938. Its endeavours to negotiate an agreement with the German Economic Ministry were cut short by a telegram from the British Government, withdrawing government support because of the German invasion of Czechoslovakia. This was symbolic of the political and economic uncertainty which hung over Europe in the 1920s and 1930s; it helps explain the lack of long-term plans among British managers.

OLD INDUSTRIES

The Textile Industry

Cotton

War produced a marked drop in imports of cotton manufactured goods, and a great increase in orders from the armed forces. Then the decline in allocation of shipping to cotton limited exports, and of course the sinking of merchant shipping limited raw cotton supplies. Some cotton manufacturers introduced short-time working or sacked employees. When their efforts to allocate raw cotton between themselves failed, the Government set up the Cotton Control Board in 1917, consisting of four employers, four trade unionists and four civil servants. Lower grade output fell as manufacturers exported only more expensive and more profitable goods. The Indian market was all but lost.

Immediately the war ended, this changed. Demand temporarily soared: so did profits. Output, though, did not rise to its 1913 level because people had left the industry and shift-working was resisted. On some processes, profits rose from ½d (0.25p) per lb of yard spun to 1/6 (12.5p) temporarily. New spindles (mostly mules) and looms were installed, and firms increased their loans from banks and issued shares which were not fully paid up. Cotton firms changed hands for large sums. Then the boom broke: the collapse was largely in exports. US firms suffered greatly. Receiving countries (especially India) put up tariffs, and increased output. But while, after about 1924, cheap Japanese exports picked up, British exports of cotton cloth fell by nearly a fifth from 1924 to 1929, though yarn — at a reduced price — stayed steady in quantity until 1929. In 1934–38 yarn exports were half their prewar quantity. British exports of cotton cloth, once 65 per cent of world cotton cloth trade, were 44 per cent in 1928–29, and 26 per cent in 1937–38. But the value of exported cotton cloth in 1937 was £67 million, compared with £126 million in 1913. Home demand remained more or less constant and was filled almost entirely by British products.

The postwar refinancing had included mergers which in theory led the way to the closure of less profitable units, and efficiency gains. In practice, little was done. Cotton manufacturers were great individualists: none wanted to close. Short-time working for alternate weeks so that unemployment benefit could be claimed by employees in the weeks between was the norm. The assumption that sooner or later export markets would recover was only slowly abandoned. In 1929, a new large-scale amalgamation took place: the Lancashire Cotton Corporation (LCC), helped financially by the Bank of England via the BIDC. Since it ultimately included almost 100 firms, it was untypical of the mergers of its time, and its object and achievement were to reduce the number of spindles in its firms by almost a half in ten years, thereby allowing surviving firms a share of the market. Perhaps this would have happened automatically if profitless firms had gone bankrupt. But profit and efficiency were not the same thing. Some of the most technically efficient firms were those which, because of rash finance in

1920, were most in debt. The LCC covered less than one fifth of the spinning capacity of the industry; it eventually paid a dividend in 1939. Spinning firms outside it obtained powers to compel firms into a self-financing scheme through the Surplus Spindles Act of 1936, which empowered the Spindles Board to take 10 million more spindles out of production, with compensation paid from a levy on spindles left running. A further act of 1939 was not used because of World War II. The much smaller Egyptian cotton section of the industry found salvation by forming Combined Egyptian Mills in 1929, a quite efficient amalgamation.

Cotton, once the flag bearer of free trade, wanted and needed government help. By 1939, four fifths of cotton goods were used in the country which produced them, often behind tariff barriers impenetrable to Britain, but not always to Japan with its lower labour costs (less than a third of British wages). More productive machinery simply did not pay in weaving, given the cost and skill of labour, the long life of existing looms and the cost of fully automatic ones. Only 3 per cent of British cotton looms were fully automatic by 1937. New methods in the pre-spinning drawing out of cotton fibres using cheaper yarns produced fibre that was soft yet strong enough to use in ring-spindles. Other technical developments reinforced the advantages of rings; and after 1930 virtually no new mule spindles were installed in Britain. But this did not save overseas markets. Alternatives were much longer runs of standard cloth, as in the USA, or lower quality cloth as in Japan. Both suggestions should be tested by asking what effect on existing markets and reputation they would have had if adopted. The 'more looms' idea was tried (see next chapter).

Some Lancashire firms diversified: cotton for lining rubber tyres, and cord for strengthening them; cotton backing for 'leathercloth' for car upholstery and kitchen table covers; and the use of rayon fibres with cotton were successful new ventures. The 1914–18 war had telescoped an inevitable change in exports; the postwar 'boom' had led to quite wrong expectations. By 1937, unemployment in the industry was still 11.5 per cent, and yet 150,000 had left the industry since 1932. A valuable export had been lost, and Lancashire's demand for manufactured goods and much else declined.

Woollen and Worsted

From 1914–18, output of these industries was almost static. Exports fell, but orders from the armed forces kept firms busy. The main problem was securing imports of raw wool in war conditions. The postwar boom saw profits, output, exports and employment rise. But unlike the cotton industry, firms in woollen and worsted did not change hands at over-high prices and so did not become burdened with debt and interest charges. The fortunes of different parts of the industries varied: the most changeable demand was that for finished woollen and worsted cloth. The 'flapper' fashion of the 1920s for shorter, straighter skirts was adopted by women of all ages and skirts, dresses and coats then used much less cloth, so demand fell. There was also a fashion for lighter cloth, largely imported. Abroad, European countries developed their own industries and bought less British cloth. There were new sales in the Far East, however, as some Chinese and Japanese began to adopt European clothing. Demand for blankets and knitwear (including socks and stockings) also fluctuated but overall, to 1924, the industries suffered far less than the other old staples.

The return to the gold standard at a high dollar equivalent then undoubtedly dealt something of a blow to exports of woollen and worsted cloth. They fell from 229 million square yards in 1924 to 186 million in 1925 and, as the Depression of 1929–33 was reached and rivals expanded production, fell further to 86 million by 1932. Nevertheless, there were compensations. There was a steady domestic demand for new wool carpets for the millions of new front rooms and, after a special tariff on imports was imposed in late 1931 for five months, a marked drop in woollen imports. When the tariff was reduced (to 10 per cent on yarn and 20 per cent on cloth), imports remained small. Britain increased exports of all types of woollen goods in the thirties and increased its share of a smaller world market, unlike its principal rivals, France and Germany. Employment in the industries had fallen from 276,400 in 1924 to 230,300 in 1930, but rose to 249,900 in 1937.

The equipment with which the goods were produced was old, apparently indestructible and, by newer standards, labour-intensive. Even by 1947, 97 per cent of the spindles used were mules, and only 6 per cent of looms were automatic (though they were power-assisted). Yet the industry responded well to changing tastes and changing markets.

The Iron and Steel Industries

The war increased demand for iron and steel and, at the same time, reduced imports of low-phosphorous iron ore, used in many British furnaces. The new Ministry of Munitions ensured that the army accepted 'basic' steel, which used high-phosphorous ores found in Britain, and steelmakers were persuaded that scrap metal inputs were reliable. The latter rose from 15 per cent of input in 1913 to 50 per cent in 1918. Old pig-iron furnaces were restored to use; new steel furnaces were added to existing plant. The government controlled and subsidised steel prices, to prevent higher prices being passed on and to forestall wage claims from men whose wages were related to the price of steel. Output rose from 7.7 million tons to 9.5 million tons per annum, 1914–18.

Iron and steel shared in the post-war boom, freed from all government controls. Some steelmakers had been apprehensive about a return of dumping on the British market by foreign makers, but the Belgian steel industry had been ruined by war, much of the French industry destroyed, and the German industry handicapped by the loss of Alsace and strikes in the Ruhr. In the boom, some British firms amalgamated and raised more capital. However, they did not integrate their works: too often metal had to be cooled and transported between processes. Too often the piecemeal wartime additions, made when cost was secondary to output, proved expensive to operate. Demand collapsed after June 1920, costs mattered more, and imports began to reassert themselves. Some of the companies with the newest equipment found it hard to keep up interest payments; over the next two decades, export markets proved unpredictable and home markets fluctuated, much influenced by demands from shipbuilding.

Belgian steel, made in the newest plant and produced by men who worked at half the British wages, reappeared and was sold at a third below British prices, for a limited range of semi-finished products. German steel was cheaper than British until lower demand limited its scale of production after 1928. German wages were about a third lower than British. Were higher British wages no longer justified by superior British skills? Undoubtedly skills were needed to get the maximum out of older plant; in any case the highest paid workers agreed to wage cuts twice (see next chapter). Belgian workers went on strike in the 1930s in an attempt to increase their pay and disrupted production. Britain avoided this. US steel workers were paid twice as much as British, but their cheaper ore, larger furnaces and more mechanised equipment cut costs. What mattered was not the absolute or relative wage paid, but whether the total mixture of wages, raw materials, fuel, and capital equipment produced steel which was competitive in price, quality for quality. High-quality British steel met this test and was therefore preferred, especially for shipbuilding.

However, it was possible that more lower-grade steel could have been sold had costs been lower. Apart from wage reductions, British steel managers pursued some other cost savers, although slowly. They at last financed research into better fuel efficiency in existing furnaces, and fuel consumption was cut by over a third, 1923–1938. Two firms — Colvilles in Scotland on the Clyde and Stewart and Lloyd in new works in Corby — used integrated processes, and smaller firms did likewise. Stewart and Lloyd used the high-phosporous ores of the region (steelmasters had often been criticised for not tackling these ores energetically, though they were not easy to use economically). But furnaces of the size of German or American ones were not built. Had they been, other furnaces would have been scrapped. Steel masters were reluctant to lose their independence, and neither they nor the government wanted to add to unemployment as fewer, larger, more modern furnaces would have done.

In fact, although for a short time unemployment in the industry was 42 per cent, British steel output during the Depression suffered less than its competitors. Only a few individual firms got help from the BIDC. 1932 was the worst year, with 46 per cent unemployment in the industry. Steel was helped by the 10 per cent import duty of 1931, raised to 33⅓ per cent after a special case was presented to IDAC. IDAC was sympathetic to the industry, but not to the in-fighting between its sections. Eventually they combined to form the British Iron and Steel Federation (BISF), and IDAC was persuaded by it to raise tariffs to 50 per cent for a short time, to give the British industry some muscle in its application to rejoin the iron and steel cartel, abandoned in the Depression and reformed in 1933. The German industry was government-controlled and geared to employment and future defence use rather than to exports. Belgium and France were the important rivals of the 1930s: the USA exported only a small percentage of its vastly greater output.

The biggest change in production technique was in strip mills which produced finer-gauge steel wanted for car bodies, and shiny tinplate for the canning industry. Metal was rolled in a continuous strip in sheds which were at least three quarters of a mile long. The process was well established in the USA, but after the uncertain 1920s, British firms hesitated to invest in it. Britain in 1929 had provided one third of world production and two thirds of world exports of tinplate, much of it from 67 small works in South Wales. Richard Thomas, which wanted to adopt the new process, was pressured by the government to build its new works and mill in Ebbw Vale. This was not the ideal site, but it was an area of high unemployment. Quarrels and compromises surrounded its construction, and of course it put the small old mills out of production. But of its kind, it was a well-equipped plant, finished with money from BISF.

Total output of steel grew in the 1930s to 13.75 million tons, double the 1913 figure, helped by the tariffs and by its continued reputation for quality. It was produced largely for the home market, though sometimes exported. A lower price in the 1920s might have helped it somewhat, but in the 1930s, price was not the main consideration. It rose to over 9 per cent of total world output, expanding to meet rearmament demands. There were some up-to-date plants, but it was not an up-to-date industry.

Shipbuilding

In 1914, Britain had the largest navy and the largest merchant fleet in the world, both built in British yards. Those yards had also built well over half the world's shipping. War, particularly submarine warfare, destroyed shipping faster than it could be replaced. In 1919 and 1920, orders flowed to British yards from British and foreign shipping companies. The 200,000 insured shipyard workers of 1914 were increased to 350,000, and shipbuilding capacity enlarged by over 30 per cent.

Then demand for shipping space fell as the postwar boom declined, and so orders for new ships declined. In any case, war damage to existing ships was repaired so that they could be used again, and shipping schedules which had been dislocated by the war were resumed, making better use of existing vessels. Further, many foreign countries had enlarged their shipbuilding capacity during the war in the effort to keep up with losses, and these yards were still working, often subsidised by their governments who wanted them to survive for social and strategic reasons. British naval orders shrank, since the German fleet was no longer a threat, and agreements with the USA and Japan limited the size of Britain's fleet. The British shipbuilding industry was responsible for none of these things, yet it has frequently been criticised for the fall in orders. An even bigger fall occurred with the Depression of 1929–32 when unemployment in the industry reached 60 per cent. Could the industry have done more to help itself?

British shipbuilding supremacy had been built on steam technology, still being improved in the 1920s. Scandinavian countries, without Britain's coal resources, had the incentive to develop and build new and better diesel engines and motor vessels, and were technically ahead of British yards in some aspects of this type of motive power. Nevertheless, in 1927–30 Britain built 41 per cent of all new motor ships, and 65 per cent of all new steamships, and subsequently rather more of the former: not a bad performance (see Table 5.1).

Then it was said that British managers, unions and shipowners were slow to accept welding instead of riveting to hold the steel plates of hulls together. But there were real doubts about the safety of welding,

Table 5.1 *Value and Percentage of British Shipbuilding Output by Type of Product*

Value of Output (£m.)	*1907*	*1912*	*1924*	*1930*	*1935*
Merchant ships: steam	19.6	} 30.0	28.6	20.9	8.3
Merchant ships: motor			7.0	18.7	8.4
Warships	3.5	6.8	0.6	4.0	6.4
Other construction	1.4	1.8	1.1	3.0	0.9
Repair	8.4	9.7	17.4	16.1	11.8
Total	32.9	48.3	54.7	62.7	35.8
Percentage of output	*1907*	*1912*	*1924*	*1930*	*1935*
Merchant ships: steam	59.6	62.1	52.3	33.3	23.2
Merchant ships: motor			12.8	29.8	23.5
Warships	10.6	14.1	1.1	6.4	17.9
Other construction	4.3	3.7	2.0	4.8	2.5
Repair	25.5	20.1	31.8	25.7	32.9
Total	100.0	100.0	100.0	100.0	100.0

Source: Buxton, N.K. and Aldcroft, D.H. (eds.) (1979) *British Industry Between the Wars*, p. 81, Scolar Press.

particularly on bigger ships. When the USA first used the technique in World War II on its Liberty ships, they literally cracked in two; not enough had been known about the rigidity of riveted hulls. Some caution had been justified; but critics held that it was too automatic.

Dutch yards were improving the outlay and design of their expanding yards; British yards made little change. Yet had they done so, it would have meant more capital outlay which would have bankrupted yards when orders shrank. On the positive side, once it paid to do so, British yards did away with steam and hydraulic power and converted to electricity. But there was room for more electric power, and for effective change in the organisation of work in the yards to eliminate narrow definitions of skill which prevented the most economical use of manpower. Such a reorganisation might well have reduced costs and increased profits, or perhaps have enabled prices to be reduced, so tempting more orders, though these were bound to be limited. But it would also have cut the numbers employed, as owners, managers, and employees alike were well aware. Already faced with resistance over possible money wage cuts, employers backed away from causing strikes over skill use.

Yet those skills did produce ships which retained their reputation for quality. It is difficult to make comparisons about price, since quality and design differed but, taking those two things into account, British yards competed well on price, in spite of obvious inefficiencies.

British shipping lines, when they ordered any ships, ordered British ships. Eventually public money was given to enable work to restart on the huge Queen Mary. It had been suspended when the Cunard Line's traffic and finances had dropped abysmally. A small government subsidy was paid for British-built tramp steamers in 1934, and in the same year all major British shipyards joined the National Shipbuilders Security Ltd. It bought and dismantled about one third of British shipyard capacity, paying for this by a 1 per cent levy on new ships, and backed by a BIDC guarantee. The idea was that the remaining yards could reduce costs by specialising in a given type of ship and also by agreeing to minimum prices to prevent yards taking on unprofitable work as they had been doing. It did not work quite as intended. Revival came with upturn in world trade, and then with trebled naval orders as war loomed again. One million tons was on order in 1938. War had encouraged maximum British output; the brief postwar boom encouraged capacity expansion which was never really justified. It had always been an industry with recurrent ups and downs, but the interwar years were quite beyond experience. The unemployment of the depression years became firmly embedded in the memory of the industry, making resistance to any labour-saving equipment or practice virtually certain. This had more consequence for the future than for 1919–39.

The Engineering Industry

The diverse output of the engineering industry was reduced to fewer products during the war but demand was strong. Furniture makers and clothing manufacturers were drawn into war production of engineering products.

Machine tools were essential for this production. Prewar specialised output was expanded and imports from the United States increased. Postwar, these tools were sold off by the government, thus upsetting the British market for new tools; then fluctuating markets for heavy engineering meant that such firms were cautious about investing in new machine tools. Small machine tool firms continued to supply known customers. When demand revived in the 1930s, the British machine tool industry could not meet it in time, and so again imports were sought. British firms did not produce the more sophisticated machine tools of the USA and Germany, and marketing in Britain did not educate customers to accept a standard product, which would have had a cost advantage. The Association of British Machine Tool Manufacturers, formed during World War I to parcel out government orders, tried to encourage specialisation, although Alfred Herbert, the largest manufacturer, stayed outside the organisation. The tools themselves were of good quality and were redesigned to use tungsten carbide in their cutting edges; the value of output, £3.5 million in 1924, was £8.3 million in 1935, and rising. About half was exported, until rearmament in 1937 and 1938 strengthened home demand. Pre-1914, Britain had supplied about 13 per cent of world trade in machine tools; by the 1930s it had about 10 per cent of this unpredictable market. Britain was not in the forefront of new design or forceful marketing, and lacked trained design teams.

Between 1924 and 1937 the output of mechanical engineering increased at about 1.9 per cent per annum — that is, more slowly than manufacturing industry as a whole and more slowly than world demand. It experienced great fluctuations in employment, though some of the apparent surge and fall in 1919–21 may be due to overlap between the returning men and the departure of 'diluted' labour. On average, it accounted for about 9 per cent of manufacturing employment and 6 per cent of manufacturing output and this reflects the specialist hand-built nature of many British engineering products.

The government built over 200 arms factories between 1914 and 1918, managing some itself, putting others under the control of boards of local businessmen, and the remainder out to private firms which ran them on an agency basis. Many of the 200 were large, well-automated, and designed to maximise the use of quickly-trained workers. Table 5.2 tells the story. The output of machine guns also increased and by 1918 armoured tanks were produced in growing quantities.

Postwar, orders and output fell abruptly. In the early 1930s the view

Table 5.2 *British Production of Rifles, 1914–1918*

Rifles	*Total*	*Contractors*	*State Factories*
Aug-Dec 1914	120,093	68,517	51,567
1915	613,461	341,605	271,856
1916	952,928	534,645	418,283
1917	1,205,672	565,559	640,330
1918	1,062,072	435,742	626,330
Total	3,954,226	1,946,068	2,008,158

State factories include both Royal Ordnance Factories and government-owned factories, whatever their management.

Source: *History of the Ministry of Munitions*, Vol. VIII, Part 1, p. 91.

developed that World War I and other wars had been caused by armourers of all nationalities, 'The Merchants of Death'. In Britain the government appointed the Royal Commission on the Private Manufacture of and Trading in Arms in 1935. By the time it reported in 1936 (against the prohibition of private armament factories, though in favour of their greater control) events in Germany had persuaded the government, reluctantly, to rearm, and this caused output to rise again.

The initial British postwar boom in cotton meant a demand for textile machinery of known designs; thereafter some home orders were for non-standard machines, to produce a different type of weave or texture. Decline in British cotton goods then led to reduced demand. The bigger companies, especially Mather & Platt and Platt Brothers, sold more machinery abroad than in Britain; the proportion fluctuated from year to year as successive countries set up their own textile industries. From 1923–9, Britain sold 49 per cent of the total exports of the six largest exporting countries, although by 1938 Germany's exports exceeded Britain's and Japan made a cheaper, basic product. German knitting and finishing machines were often superior to British, but British spinning and weaving equipment was still first class. Except for hosiery and knitting machines, the British home market, helped by tariffs, remained the preserve of British manufacturers.

In the 1920s and 1930s, all four British railway companies experienced nil or low return on capital. Their works, while producing prestigious new locomotives, were much occupied with repairs. They depended heavily and quite successfully on exports. British-built steam engines lasted a very long time, and so replacement orders were limited. There was some demand for new locomotives for electrification of some British suburban railways (the government lent money for this), and some experimentation with diesel engines; foreign locomotives and rolling stock made virtually no sales in Britain. Market retention was commendable, at home and abroad.

Small engineering firms with low overheads sometimes did better than large ones in the Depression: small makers of cranes and mining machinery could find a flow of orders just sufficient to keep a team of men working on one or two pieces of heavy equipment at a time. The real difficulty lay in the combination of low average firm size (150 workers or less was the norm) with low profits. This led to a lack of investment in applied research which could develop new engineering products. Though there were laboratories in some bigger firms, and some DSIR support, these were directed largely to improving existing products. Nevertheless, imports were negligible.

NEW INDUSTRIES

The Engineering Industry

Light Engineering

The growth of office work, government and non-government, and the employment of girls and women as typists and secretaries, created a growing market for office equipment. The principal British firm, the Imperial Typewriter Company, expanded output. Two leading American producers, Underwood and Remington, established assembly factories in Britain but imported parts from the USA.

New items for leisure included small cameras. The simple but adequate Box Brownie, produced by Kodak Ltd, the British arm of the large US firm Eastman Kodak, was made in Britain from the 1920s, first from imported parts but then all British-made. New items for the home included tin-openers, to deal with increased consumption of canned foods. Small components for domestic appliances, cars, aircraft, and electrical equipment were made in light engineering firms. Sewing machines kept their home outlets but lost some export markets.

Motor Vehicles

The production of private cars virtually ceased, 1914–18. Postwar, demand surged. Austin, Morris and Ford were the three largest manufacturers in Britain, each making comparatively inexpensive cars. Ford, abandoning its Model T in the 1920s, produced a car with an improved gearbox, higher-powered and slightly more expensive than the very economical Austin 7. Morris cars were midway in power and price and had the differentiation between models that British car owners wanted. These firms, plus the three medium-sized ones of Standard Motors, Vauxhall Motors (US-owned after 1925) and the Rootes Group, shared the market with 90 other firms in 1922. By 1939, the number had dropped to 26, and many of the smaller firms had financial links with the big six.

Ford moved from Manchester to Dagenham in Essex in 1931, but most car factories were in the Midlands. Trucks and buses, with greater reliance on traditional heavy engineering, were made in Lancashire (for example Leyland Trucks, which evolved out of a steam lawn-roller business), and North Cheshire, and the outskirts of London. All large car firms adopted Ford's idea of assembly lines postwar, and the making of engines and body sections was to a degree automated. This led to a fall in the cost of cars and Morris led these price cuts. Austin two-seaters cost £225 in 1922 but £110 in 1932; then increased, as the cost of inputs rose, to £122 in 1939.

Production of motor vehicles grew as Table 5.3 shows. The decline to 1931 was followed by a revival, and the loss of production was small compared with the 70 per cent fall in US output. By 1937 Britain, producing only one tenth of the cars made in the USA, was the world's second largest car manufacturer. Increase in output meant more inputs: the British steel industry was extremely thankful for the industry's orders. Lucas in 1926 acquired both the other major electrical component firms and had a virtual monopoly in starting, lighting, and electrical equipment.

Table 5.3 *UK Production of Motor Vehicles (000s)*

Year	*Cars*	*Commercial vehicles*	*Agricultural tractors*
1913	34		—
1922	73		22
1923	71	24	n.a.
1925	132	35	n.a.
1929	182	56	10
1930	170	67	15
1931	159	67	4
1932	171	61	3
1933	221	66	3
1934	257	86	4
1935	312	92	9
1936	354	108	13
1937	390	118	18
1938	341	104	10
1939	305	9	15

Source: Silberston, A. (1958) 'The motor industry', in D. Burns (ed) *The Structure of British Industry: A Symposium*, Vol. II, p. 4, Cambridge University Press.

There was no subsidy for the motor industry, but first the McKenna duties and then the 'strategic industries' tariff protected it. Vehicles paid a horse-power tax. Therefore, foreign cars which on average had more powerful engines were taxed more heavily. A 1928 tax on petrol, increased in 1931, reinforced demand for low hp cars. In the 1930s, the dollar's fall in value led to a price reduction in imports, and a 25 per cent reduction in hp tax in 1935 gave some incentive to larger-engined cars. But more British buyers were economy-conscious than in the 1920s, and lower hp British cars retained their popularity. Smaller, cheaper German and Italian cars were imported, but before claims in 1937 that these had been dumped were investigated, the imports fell. About 87 per cent of cars in Britain were British-made.

British shares of the changeable international market fluctuated. When sterling devalued in 1931, British cars sold more cheaply abroad for a time and exports benefited. Their low hp, low running cost qualities had appeal. Empire/Commonwealth markets usually took well over 70 per cent of British car exports, and both before and after the Ottawa Agreements gave British vehicles tariff preference. But when importing countries put higher duties on fully assembled cars than on imported chassis, US companies were quicker than British to set up subsidiaries abroad to undertake this assembly. In any case, their cars' springing was better suited to the rough roads of the outback.

Ford's British models did not sell well in Italy, France or Germany. Its Model T was eventually priced at £100 for Britain. This increased sales, and yet the car was outsold by dearer British cars with more comfort and more differentiation. Therefore, Ford's hoped-for long production runs at Dagenham did not materialise.

In the 1920s, horse-drawn commercial vehicles were virtually ousted from towns and cities by light vans and lorries. By the 1930s, buses and trolley buses were replacing trams. Heavy lorries increased sales after the production of more efficient diesel engines in 1933. However, both passenger- and goods-carrying by road was, from 1930, subject to considerable government regulation — not least to shield the railway system from competition. More legislation in 1933 (which put higher taxes on heavier vehicles) was the stimulus for lighter trucks with better payloads. One-ton Morris trucks were popular, if noisy. Imports of heavyweight commercial vehicles were never more than 5 per cent of British sales. Light-weight imported trucks, almost 25 per cent of the home market, decreased once Ford and General Motors set up commercial vehicle manufacture in Britain. The destination and the percentage exported of commercial vehicles was similar to that of cars. Most were high in price and quality, and were specialist vehicles in the best, but difficult-to-expand, tradition.

The motor industry grew faster than any other new industry between the wars. Employment in it was perhaps 9 or 10 per cent of all manufacturing employment by 1938. It stimulated demand not only in inputs, but also in service industries — petrol pump attendants, garage mechanics, tax officers, insurance agents — and in construction — roadmakers, and garage builders, commercial and domestic. The motor vehicle industry, imitative rather than innovative, was an enormous asset to the economy.

Electrical Engineering

This industry developed new products — radios, cookers, vacuum cleaners and fridges. In 1907 it was responsible for 1.4 per cent of net UK output; by 1935 it provided 4.9 per cent, and employed 367,000 workers by 1937.

In 1926, the Central Electricity Board was set up by an act of parliament to build a national grid, and to buy and sell surplus electric power from selected stations among the hundreds owned and run by local authorities and private companies. By 1939 the national grid enabled peak demand in one area to be satisfied by surplus electricity from another, thus lowering costs. The amount of coal needed to generate a given output of electricity fell markedly, and the price fell too. The detailed distribution of electricity was in the hands of local authorities and the companies. By 1939, over 60 per cent of all houses had electricity, and industry multiplied its use tenfold between 1921 and 1938. This meant a vast increase both in the demand for generating equipment and distribution cables, and for electric motors and domestic appliances.

During the war, the German-owned Siemens Dynamo Works was confiscated by the British Government; afterwards it was sold to a new company, English Electric, an amalgamation of two other British companies. Associated Electrical Industries, except for a short interval, owned Metropolitan Vickers; in the merger wave of the twenties it also acquired British Thomson Houston and Edison Swan, and was itself a subsidiary of the American General Electric Company. 'MetroVick' was unusual in the great attention it paid to research and to the training of apprentices; other firms relied on US — and sometimes German — know-how. By the later 1930s, the turbines and generators of the larger, newer British power stations were comparable in size to those in Germany and the USA.

Demand for cables, both for the national grid (over 4,000 miles) and for the distribution system replaced prewar orders for telegraph cables. Cable manufacturers had advanced research teams and in 1920 they joined the government and the manufacturers of generating equipment to form the British Electrical and Allied Industries Research Association. For research costing £4,000, the Association showed that previous insulation of underground cables was excessive; cheaper cables could take the same load, saving £250,000 per annum. The Institute of Electrical Engineers became the centre for technical debate on the design of distribution systems. In this instance, British applied engineering was doing well. Profits were maintained by a powerful trade association of producers which covered 90 per cent of cable output, controlled prices and rotated contracts between them, especially as demand fell between 1929 and 1933. It recovered again with the expansion of power lines after 1935.

In 1914, most light bulbs were imported. The war fostered a home-grown industry. Between 1925 and 1935 production trebled, while the price of bulbs halved through increases in the scale and efficiency of manufacture. Large firms dominated, producing everything from transformers to torch bulbs. They were linked to world-wide patent-pooling agreements, and price-fixing arrangements on the home market. Aided by protection from 1921 onwards, the British-based industry resisted imports and price cuts. Cheaper independent producers in Britain forced down prices to some extent; but later imported Japanese bulbs were cheaper still, if not as durable. British manufacturers agreed among themselves to make cheaper bulbs — but to give them limited durability.

After electric light, the largest number of electric articles was radios, initially worked on a 'crystal', with earphones. In 1922 radio manufacturers formed the British Broadcasting Corporation, thus helping create demand for radios. It was given a government monopoly and financed by licence fees and payment of royalties on each set sold by the manufacturers. In 1926 the government corporation, the British Broadcasting Commission, took over the company's monopoly. By 1939, there were about 9,000,000 radio sets in Britain. The British radio industry was protected from competition by tariffs of 33⅓ per cent and by collective agreements between manufacturers. Valve manufacture for radios had been a wartime development. Postwar, the market leader, Mullard, wholly-owned by the Dutch firm Philips of Eindhoven, from 1927 used technology invented in Britain and developed in Holland. Nevertheless, the British industry was not technologically stagnant, as the success of advanced valve development for radar carried out by MetroVick (transmitters) and A.C. Cossor (receivers) in the late 1930s showed.

By 1939, of houses installed with electricity, 77 per cent had irons, less than 40 per cent had vacuum cleaners, 27 per cent had electric fires, 16 per cent had electric kettles, 14 per cent had electric cookers. Electric water heaters, washing machines or refrigerators were in less than 5 per cent of households. Many smaller British companies tried to establish themselves, but the quality of wholly British design varied. Research into television in Britain took place at the laboratories of EMI (Electrical and Musical Industries), formed in 1931 by the merger of British and US gramophone, radio and record producers. Marconi (fully British) did work on the development of transmitters. The EMI/Marconi system was fully electronic and was chosen despite pressure for the BBC to use the earlier 'British' Baird system. EMI had licensing and know-how agreements with both RCA (USA) and Telefunken (Germany) but this time a British 'new' industry contributed at least as much as it received to technical and potential commercial development.

In 1938, 5 per cent of all those engaged in manufacturing were in the electrical engineering industry. The annual increase in simple labour productivity was about 2.3 per cent per annum — not spectacular. However, capital input was very low in some sectors. Had assembly been more automated, employment might well have been lower and productivity and output higher. As it was, British products (aided by protection) dominated the home market and, in real terms, exports rose 45 per cent above their 1913 level by 1937, a creditable achievement. What was lacking was inventive research and design in household items.

Aircraft

Aircraft were originally playthings for the rich and adventurous. Commercial manufacture of aircraft was pioneered in Britain. Short Brothers acquired licences to use the Wright Brothers' patents in 1908, and began manufacture in the same year. Other companies followed, and the state-owned Royal Aircraft Factory manufactured airframes and aero engines for the army. In 1914, the British army and navy had 113 aircraft and 6 airships between them, to use for reconnaissance. By 1918, the newly-formed Royal Air Force had 3,300 aircraft in the first line and 103 airships, with thousands more in reserve. Over 50,000 aircraft were manufactured in Britain, 1914–18.

After some disastrous engine designs, the Royal Aircraft Factory ceased production. Private manufacturers were strictly controlled by the Ministry of Munitions, which supplied the capital. By November 1918, Britain had by far the largest aircraft industry in the world. The growth of the industry may be seen from the figures set out in Table 5.4.

Production of large numbers of aircraft revealed important shortages. Cellulose acetate, 'dope', was needed in quantity to treat the fabric with which the (wooden) aircraft frames were covered. There was no British supply, so the Government asked a Swiss entrepreneur to establish a factory in Britain in return for guaranteed contracts and a monopoly position in the British market. This led to a financial scandal, but also to the eventual establishment of a major rayon firm, British

Table 5.4 *Growth of Aircraft Production, 1916–1918*

Date	*Number of firms*	*% of 'dilution' of labour*	*Number of workers*			
			Total	*Men*	*Women*	*Boys*
Aug 1916	491	31.7	60,073	40,974	12,615	6,484
Nov 1917	771	40.1	173,969	104,102	52,734	17,133
Oct 1918	1,529	46.0	347,112	187,526	126,544	33,042

Average monthly output, 1917: airframes 1,229 engines 980
Average monthly output, 1918: airframes 2,668 engines 1,841

Source: Jones, H.A. (1937) *The War in the Air*, App. XXXIII, The Clarendon Press.

Celanese Ltd. Again, almost all magnetos used in Britain in 1914 were made by a German firm, Bosch. A Bosch magneto from a shot-down Zeppelin was taken to pieces and a similar magneto devised and produced by Westinghouse. Lucas bought a small British firm and made 50 per cent of wartime magnetos.

In peacetime orders for military aircraft were not maintained. Civil aircraft firms faced a hazardous market because of the large postwar stock of aircraft and spares, cheaply sold off to the civil market until 1930, and the 'airlines' themselves were small-scale. The government subsidised one large firm, formed by a merger of the four main airlines, creating Imperial Airways in 1924. This meant a small but guaranteed market reserved for British aircraft makers. On the whole, British civil aircraft were technically behind foreign, and particularly American, aircraft which had been carefully developed for a larger domestic market. There was practically no export market for large British civil aircraft and all export orders fluctuated. Most successful was the de Havilland Gipsy Moth, a light aircraft first produced in 1927. By 1937, of the 1,682 civil aircraft registered in the United Kingdom, 724 were de Havilland planes; de Havilland's annual exports fluctuated from one tenth to one half of their output.

The Air Ministry wanted a diversity of designs and designers. It gave aero engine orders to four firms and airframe orders to a larger number. Good work was done on aspects of basic research and aspects of design at the National Physical Laboratory and the Royal Aircraft Establishment, though much technical improvement originated abroad. Military ordering of aircraft sank to 445 in 1932, when disarmament talks were in progress. In 1935 it was decided to re-equip the Royal Air Force. Eight thousand aeroplanes were ordered in 1936; two years later only 4,500 had been delivered. Many aircraft factories were little more than large experimental workshops, with no experience of large-scale production. Even the larger firms needed retooling. Therefore the government built large factories called 'shadow factories' which could be managed by car firms or by larger aircraft firms, both of which had some appropriate experience.

In 1930, only about 21,000 were employed in the industry; in 1935 there were 35,000, and by 1938 120,000. New developments included the replacement of open-cockpit, wooden biplanes by streamlined all-metal enclosed monoplanes with a retractable undercarriage, although neither the designs nor the tools to make them had been perfected, and there were shortages of skilled engineers and draughtsmen. Good new designs such as the Spitfire emerged. When they went into production, the increasing sophistication of the product contrasted with the skilled-craftsman tradition which assembled it. The development of civil aircraft was shelved.

The Chemical Industry

The chemical industry accounted for only 1.1 per cent of gross national product in 1907, and 2.1 per cent in 1935. Nevertheless, its role was strategically crucial in war and important as a new and growing industry in peace. War cut off supplies of dyestuffs from Germany; but dyes were urgently needed for thousands of uniforms. Small British dyestuffs firms tried to increase production, but needed coordinating, and so the government and the dyestuffs users established a joint company (British Dyestuffs) to take them over and finance research in industry and the universities on a scale unknown hitherto. Britain became almost self-sufficient in a range of basic dyes, but the company put much of its efforts into manufacturing a chemical for explosives. With peace, the German firms, with cheaper and more varied dyes, reappeared and the firm was in trouble. The government restructured it; then the slump of 1920 hit it and, although it was given tariff protection, it made only a partial recovery. Its research was cut back, but it did make an important discovery — a dye which could dye the new man-made fabric, cellulose acetate. It tried for years to come to an understanding with the German dye cartel, but this was vetoed by the government since it would have lost some of its markets for ever. What was important was not just dyestuffs themselves, but the growing range of fine chemicals which its materials and processes could develop. Ultimately, British Dyestuffs (BDC) was made part of Imperial Chemical Industries (ICI), its research was expanded, and its work formed the basis for ICI's organic chemicals in the 1940s and 1950s.

The war, of course, vastly increased the demand for explosives. The British Nobel Company, separated from its German holding company, responded. Its products included cordite for a propellant for guns, though this was also made by smaller factories and government works. When the war ended, the British Nobel Company merged all British explosive firms into one under a holding company eventually called Nobel Industries, presided over by a vehement Scot, Sir Harry McGowan. From 1915, Brunner Mond, UAC and Castner Kellner also made explosives: Brunner Mond devised means of producing ammonium nitrate for one explosive and synthetic phenol for another. The company grew rapidly, according its scientists, though not its engineers, high esteem. UAC hung on, producing explosives and a variety of chemicals; and Castner Kellner, with 25 per cent of its shares bought by Brunner Mond, made poison gas (little used) from chlorine. These three firms were not part of Nobel. The latter, to compensate for the huge drop in demand for explosives when the war ended, used its profits to buy shares in other firms, especially those which supplied the motor industry. By 1920, 60 per cent of its capital was in 'civilian' firms and its largest investment was in the US General Motors. Brunner Mond used its profits to buy the whole of Castner Keller and diversified backwards into limestone and tar, making large outlays on plant and machinery. After some hesitation it bought plant at Billingham from the Ministry of Munitions. Using research done in the Ministry, it made synthetic ammonia. This in turn made nitrates, used for explosives — or, in peace, for fertilizers.

McGowan exercised tight financial control over his companies. When, in 1925, the German chemical firms consolidated themselves into one huge firm, I.G. Farben, he persuaded Brunner to combine Brunner Mond with Nobel and to add BDC and UAC. This created Imperial Chemical Industries in 1926 (BDC and UAC were consulted — and included — later!). ICI was wealthy enough to finance truly large-scale production, investing £20 million at Billingham for fertiliser plant. The aim was to supply the entire fertiliser needs of the British Empire, then at its peak in area. The domestic market was also to be important. However, by early 1929 it was apparent that world demand for fertiliser was not growing as fast as world supply and, speeded by the slump of 1929, demand dropped drastically. Half the new plant at Billingham was closed, 20,000 employees were dismissed, and wages cut (temporarily) by 5 per cent. Size, cost control and technical prowess were not enough: the market had to be anticipated.

Fortunately, ICI was big enough to stand the loss. It used some of the plant to make petrol from coal, employing some overseas patents. Produced in 1936, it was dearer than petroleum made from oil. The government agreed to protect it because of its strategic importance for Britain which was then thought to have no domestic oil supply. The depressed coalfields — and the government — were glad of this demand for coal. But output was fairly small.

ICI's divisionalised structure (rather like that of large US firms) and its operating policies were a blend of McGowan's and Mond's approaches, and an example to other British companies. By 1938, it employed 500 research scientists. It protected domestic and Empire/Commonwealth markets by agreements with large foreign firms. Since, worldwide, there was over-capacity in heavy chemicals, this at least provided a share for ICI.

Other firms developed other products. Glaxo, beginning as a New Zealand firm producing dried milk for babies, built up pharmaceutical research and products. Beecham expanded from pills and powders into a range of health care products and toiletries. There was diversity of size and product, in spite of the giant ICI, but the pre-1914 competition in heavy chemicals was much diminished.

The Synthetic Fibres and Textiles Industry

In 1914, Courtaulds dominated British rayon production. The 1914–18 war prevented rival European firms from setting up in the USA and Britain, and in 1917 Courtaulds bought a sequestered German firm in Flint from the British Government. However, it had a new competitor — British Celanese — which used cellulose acetate to make rayon (see pp. 99–100). It was dearer and for some years harder to dye than Courtauld's viscose rayon, but it was stronger and could be spun finer. Both firms produced their own fabrics. In the 1920s, these were largely made of knitted filaments. British Celanese ran an effective advertising campaign, and Courtaulds responded; the latter depicted slim, elegant, straight-haired women; the former leaned towards curves and curls. Underwear, stockings, and the linings of men's suits were made of rayon.

Courtaulds in the 1930s remained much the bigger firm, but it neglected research; it did, though, produce a new item, staple fibre (patented as 'Fibro') — rayon filament cut into shorter lengths which could be used with cotton or wool. Its importance had not been seen at first, but by 1939 one third of total rayon output was of staple fibre, cheaper because easier to handle than filament. Courtaulds also went into cellulose acetate rayon production as the price gap between the two processes closed. In the 1930s both it and British Celanese produced a less shiny fabric, with a different set of uses, including woven dress material. British Celanese, proud of its chemical research, used its basic material to make opaque celluloid, transparent cellulose sheeting (used in windows of cars with collapsible roofs) and cellophane-like wrapping 'paper'. Total continuous rayon filament output in Britain (i.e. both viscose and acetate) was 7 million lbs in 1913 (3.2 m. kilos, approximately), 53 million lbs in 1929, and 115 million by 1939 — 10 per cent of the world's output in the latter year. Total British rayon staple fibre was 2 million lbs in 1929 and 58 million lbs in 1939 — 5 per cent of world staple fibre. By 1935, British employment in rayon production was 35,000 and rising. Of 420,000 looms in Britain producing fabric of all kinds, 74,000 used rayon wholly or in part. In contrast to the cotton industry, the links in the chain of production were full of new equipment: it was a very capital-intensive industry.

This looks like a real success story; and if one adds that as production increased, prices came down, so that what had begun as a substitute for the luxury of silk became, at least in its later staple fibre version, only about 1½ times the price of comparable cotton, it might seem hard to criticise. But criticised it was and, as with other products, largely for its international performance: in 1925, there were more imports than exports of rayon. A 1925 tariff on rayon imports (and an excise duty of half that rate on British-produced rayon) changed that. The tariff tempted the Dutch firm, Enka, to establish a subsidiary in Britain. Courtaulds came to an agreement with Enka's parent, but brought its own prices down to prevent other firms from successfully setting up in Britain. This price cutting showed that the higher prices of the early 1920s were not a reflection of cost, but of the profitability of virtually unchallenged sales. Only nine of the postwar newcomers survived.

By 1922, US output of rayon surpassed Britain's. By 1926, Italy's did the same, and Germany's in 1929. The USA of course had a bigger home market. The Italian rayon was largely for export, its cheapness due to cheaper labour costs. But it was also the case that, in totalitarian Italy, Germany, and Japan of the 1930s, rayon production was enhanced by laws requiring a percentage of the fibre in all textiles to be produced in that country, to cut down imports of raw cotton and to some extent of wool. From 1934 to 1938, Italy was the major exporting nation. However, this is not the whole story. Courtaulds had built subsidiaries in most European countries, held a joint controlling interest in the largest Italian rayon company and owned the American Viscose Company which continued to supply a large proportion of Courtaulds' profits until the later 1930s. Therefore a small percentage of exports did not mean inattention to international sales. In 1930, a world surfeit of rayon cut back production in Italy by half and in the

USA by a quarter, but British output was steady. Increasingly, agreements between rayon companies of different countries divided up world markets. Britain's limited exports went primarily to India and Australia after the 1932 Ottawa Agreements.

Critics of the export performance failed to take note of these factors, and of the unlikelihood of Britain's continuing to supply over half the world's output as technologies spread. Courtaulds was not in the forefront of new discoveries or their application between the wars, but rayon was a thriving new industry, serving the home market, and consciously creating customers for a new product.

Optical Glass Manufacture

The British optical glass industry was composed of small family firms with poor technical and scientific resources. Attempts to establish scientific research and technical education had failed. German optical glass dominated the market, and 90 per cent of British needs were imported from Germany in 1914. The British Army either used German glass, or put small orders with small British firms, including Chance Brothers (whose main product was sheet glass), which then had little incentive to develop large-scale production. Optical glass was needed for binoculars, telescopes, gun-sights and photographic lenses.

When the Ministry of Munitions was established it put Chance Brothers under pressure to increase production tenfold, with financial aid. In return, they were guaranteed a monopoly in supplying the Government. A year later, they produced 14,200 lbs monthly. They established a research laboratory which, by the end of 1918, employed 125 people.

Mass production techniques were introduced into manufacture of the instruments. Output increased in value from £250,000 to £5,000,000 per annum. By the end of the war, Britain was able to meet all its needs for optical munitions.

Nevertheless, when government orders virtually ceased in 1919, the industry declined as foreign competition reasserted itself. In spite of protection under the 1921 Safeguarding of Industries Act, Chance's sales of optical glass reached their lowest, 3,500 lbs per month, in 1922. Three optical glass firms went bankrupt in 1926. However, the Scientific Instrument Research Association, a DSIR-sponsored body, remained in being, and did valuable work. In 1932, Chance took over the Derby Crown Glass Co. and signed an agreement with the Admiralty for an enlarged programme of research. From 1934 to 1939, with defence orders, there was a fivefold increase in the manufacture of optical glass. It was a technically sound industry, but it owed its survival to government funding and purchases.

The Food Processing Industry

The large nineteenth-century steam-powered flour mills and sugar refineries reflected the increasing size of food processing firms and their changing technology. But the postwar scale of advertising and sale of branded products was new, and from 1918 to 1939 went on apace, helped by new developments in food processing and the extension of the chains of grocery shops initially designed to sell margarine to the working classes.

During the 1914–18 war, butter was in short supply, and so margarine was offered by independent grocers who dealt mainly with the middle class. When sugar, the first foodstuff to become scarce, was available only to customers registering with a specific grocer, the local Food Control Committee frequently added butter and margarine to the foods for which registration was necessary. Middle class households got to know margarine, working people felt some entitlement to butter, whose superior nutritional values nevertheless were not fully understood. Prices of both were intermittently controlled until 1921. Since butter was only 2d (.83p) more per lb than margarine, the competition among margarine producers to retain and, if possible, extend their share of the market was intense, and they bought and managed chains like Maypole, Meadow, Home and Colonial, International. Eventually, Lever Brothers moved from their soap empire into margarine manufacture, using some of the same ingredients. They merged with other amalgamated firms in 1929 to form Unilever, the largest manufacturing employer in Britain and the largest manufacturing firm in terms of capital employed — though over half their capital and half the staff were engaged in their shops. These shops stocked a wider and wider variety of foodstuffs, much of it prepacked, though some was still measured and wrapped while the customer waited. It was necessary to own the retail outlets, since strict stock control was essential to keep products fresh in the unrefrigerated shops.

The processing of food by bottling was given a boost by a cheaper American process for making glass jars and bottles, readily taken up by Pilkington, Britain's major glass manufacturer. Canned foods were thought inferior to bottled ones. Then imports of American, Argentinian and Australian canned products, especially corned beef, were welcomed during the war, and soldiers were accustomed to tinned 'bully beef'. In 1925, the National Food Canning Council was established in Britain by the canners and the tinplate manufacturers with help from the Ministry of Agriculture. The number of food canning factories grew from 6 to 80 by 1935. Most cans were supplied by the Metal Box Company, using tinplate from South Wales. In 1929, American Can — the largest United States can-maker — entered the British market. Metal Box signed an agreement with American Can's US rival, Continental Can, giving it exclusive rights to Continental Can's machinery, patents and processes in the United Kingdom. Eventually Metal Box maufactured its own machinery. So successful was the venture that American Can withdrew from Britain, selling off its plant to Metal Box in 1931. Metal Box increased output of cans from 23 million in 1930 to 500 million in 1939, educating food companies in canning techniques and hiring out can-sealing equipment. British production of canned vegetables rose from 24,000 tons in 1920–22 to 193,000 tons in 1937–38; canned and bottled fruit rose sixfold.

While there was room for many small firms, big firms predominated, because of the economies of scale. By the end of 1938, the Co-operative Wholesale Society, Ranks, and Spillers had 65–70 per cent of the flour market; the two latter firms maintained a price- and market-fixing arrangement in conjunction with five smaller firms; Unilever dominated margarine, under a variety of brand names. Altogether, some 800,000 people were employed in food processing in factories by 1938. The variety of foodstuffs available was much enhanced and the high turnover, small profit margin philosophy was beneficial to the customer. New production processes, advertising and careful quality control established new markets and unforeseen employment in manufacturing and a service industry.

Manufacturing and Research

Wartime shortages of key materials and products convinced the Government that greatly increased state support for research was needed — hence the founding of the DSIR (p. 72). It had three duties: to establish and run research institutions such as the Fuel Research Board, to promote scientific research in universities through grants to staff and research students, and to encourage research in industry by establishing Research Associations, funded jointly by public funds and the firms in a given industry. It was expected that RAs and their industries would become so convinced of the value of research that after five years they would need no public funds. Parliament voted £1 million to start them in 1917.

By 1939 over 30 RAs were set up, but in every case had continued to exist only when DSIR money was stretched beyond five years. RAs did good work, but research-conscious firms were wary of sharing new ideas with rivals. Progressive firms set up their own laboratories. The biggest research expenditure was by ICI which, after 1926, spent more than all the RAs together. Nevertheless, Britain's research, excellent as it was in quality and the development of new products and processes in dyestuffs, films, talking pictures, and radar and television, was lacking in the quantity and application found in the USA. However, except for a few industries such as chemicals and television, there was neither the capital investment nor an adequate number of technically trained draftsmen and engineers to translate research into readily marketable products. The gap between the pure science of Oxford and Cambridge and the needs of manufacturing was bridged on occasion during the war but only infrequently thereafter. Scientists from other universities were brought into the laboratories of progressive firms, but not into the boardroom.

Conclusion

The three controversies outlined in Chapter 4 take on a new dimension when looked at in the light of the fortunes of individual industries. The first controversy concerned the return to the gold standard (see pp. 77–8). A return to gold at 10 per cent lower against the dollar would have affected industries differently. It would have made little difference to cotton textiles in their competition with Japan. With labour costs more than two thirds lower, that country would have continued competing by reducing its profit margins. Tariffs would still have shut out British cotton. The price of British ships was reasonable, quality for quality; the market for them was not elastic, and was curbed by foreign tariffs and subsidies. A lower pound might have attracted a few more orders in the 1920s, but left untouched the weakness of skill demarcation and poor yard organisation. If shipbuilders had kept up prices in foreign currencies after devaluation, and had used the increased profits to install more cranes etc., the extra overheads would have made their position worse, 1929–33. Car exports would probably have benefited from a lower pound, though it might have been difficult to expand production very quickly; engineering would have benefited. However, none of this takes account of possible new tariffs being erected against British exports, nor of other currency movements, and these considerations apply particularly to steel, where France and Belgium were eager to retain their price advantage on steel of some grades.

What is certain is that a higher pound meant lower import prices for food and raw materials. Half the nation's food was imported, and its cheapness left purchasing power in the hands of British consumers who formed the mainstay of the market for the new industries. It also meant a slightly less miserable life for the unemployed.

The second controversy (see p. 68) was about the role of new industries in the economy, and whether they were held back by 'over-commitment' to the old industries, and whether the economy itself suffered because of an over-commitment. The old industries had served the economy well, prior to the war, and during it, whatever their inefficiencies, they had been needed. Postwar, some of their workforce went to new industries; the latter therefore had only temporary skill shortages, and no shortage of unskilled workpeople. Neither were there signs that capital tied up in old industries was depriving new ones of finance. While the latter could well have been more capital-intensive, especially in the thirties, and should have used more applied science and technical and processing engineers, they preferred other options. Like most new industries in most countries, they were protected by tariffs, but part of that protection was needed not because of their own inefficiencies, or higher British real wage levels, but because of the effect of the value of the pound and the need for time to build up to a volume of production that could establish economies of scale.

The third controversy (see p. 77), about the role of government in assisting the economy, is often conducted with the benefit of hindsight; in the 1950s and 1960s, when it was thought that demand management was the answer to employment, interwar governments were rebuked for not increasing expenditure and so directly or indirectly decreasing unemployment. (In fact, in some years in the 1930s the budget was in deficit, but this was concealed by the Treasury, switching money between funds.) After the experience of high unemployment again in the 1970s and 1980s, and failed attempts to put this right by demand management, there is more awareness of the remedy's limitations. Larger, earlier government subsidies for shipbuilding, encouraging replacement of older ships, would have been useful, feeding back into demand for steel, though by themselves they would have perpetuated the industry's shortcomings. Perhaps the best help would have been to provide money for the right kind of education, and at the same time attempt to persuade firms and industries of its importance.

World War I provided ready domestic markets for the old industries but curtailed exports. The brief postwar boom encouraged exports and wrongly led firms to believe that the demand was permanent, and could be met simply by using more of the existing modes of production. The slump which followed created over-caution about capital outlay, and the Depression of 1929–33 reinforced hesitancy about innovations which in the short run would have increased unemployment and in the longer run made industries better able to withstand competition. The war gave stimulus to applied science, and to some new industries. Whatever the shortcomings of British industries, unemployment in Britain 1929–33 was only half that of the oft-admired US and German economies, and this was no mean achievement for a country so heavily dependent on world trade.

Further Reading

Allen, G.C. (1970) *British Industries and Their Organisation*, Longman.
Supple, B. (ed.) (1977) *Essays in British Business History*, The Clarendon Press.

Specific Industries (in addition to those recommended for Chapter 2)

Boswell, Jonathan S. (1983) *Business Politics in the Making: Three Steel Companies Compared*, Allen and Unwin.
Carr, J.C. and Taplin, W. (1962) *History of the British Steel Industry*, Basil Blackwell.

Advanced Reading

Capie, F. and Foreman-Peck, J. (1981) 'Comments on British protection in the 1930s', *Economic History Review*, 2nd series, Vol. XXXIV, No. 1, pp. 132–142.

Questions

(1) Compare and contrast the effect of the 1914–18 war on the cotton and steel industries.
(2) Which industries in the 1920s and 1930s made good use of scientific research? Why did more not do so?
(3) Would an increase in the general level of demand have helped the staple (i.e. 'old') industries after 1929?
(4) How useful to the economy as a whole were the 'new' manufacturing industries of this period?

6

Employees, Trade Unions, and British Manufacturing Industry

Disputes between employer and employee in the workplace were not only about hours, wages and conditions of work, as the syndicalist influence on pre-1914 strikes showed. When war loomed in 1914, trade union leaders and Labour Party members discussed whether they should endorse Britain's entry. Lack of their cooperation would have impeded war production. Even strikes over the traditional targets of wages, pay and conditions could of course have had the same effect. But by the end of August, as Germany overran Belgium and drove into France, trade union leaders and the Labour Party accepted Britain's entry and agreed that for the duration of the war there would be no strikes.

However, some trade unionists were pacifists, some belonging to the small but significant Independent Labour Party which had influence on the Clyde Workers' Committee. This committee, largely consisting of shop stewards, first organised strikes against deferment of a small pay claim (late 1914), and then in late 1915 against rent increases. These strikes spread rapidly on industrial Clydeside. In coal mining too there were strikes. The borderline between 'industrial' and 'political' strikes was virtually impossible to draw in war conditions. The Munitions of War Act (see section on trade unions and the law below) could not abolish strikes; in early 1916 government decisions to end 'trade cards', which gave exemption from conscription to members of skilled unions, and to extend dilution to non-war work, led to a successful strike, the biggest of the war.

Yet the war saw new cooperation between government and trade unions. Lloyd George, as Minister for Munitions and as Prime Minister, talked to and understood union leaders, and from 1916 made

Arthur Henderson, Labour Party leader, a member of his inner war cabinet of five. The Russian Revolution of 1917 and Russia's subsequent withdrawal from the war produced new divisions within unions and the Labour Party. Henderson favoured a negotiated peace, and many shop stewards agreed with him. But uneasy cooperation with the Government lasted until the end of the war in November 1918. In the postwar election, Labour won 59 seats; most Labour MPs were trade unionists. The Labour Party wrote specifically socialist aims into its constitution, including 'common ownership of the means of production, distribution and exchange', and the most equitable distribution of the 'full fruits of their industry' to 'workers by hand or brain' on that basis. Would trade unionists outside Parliament now concern themselves only with day-to-day industrial disputes and their members' workplace welfare, leaving wider matters to the Labour Party and its MPs? The answer was no.

Union leaders and Labour Party alike were hostile to the idea that the government might continue conscription and continue sending troops to aid anti-revolutionary fighters in Russia. The real possibility of widespread strikes made the Government back away from both thoughts. Similar opposition, with a similar strike threat, helped cause Lloyd George (still Prime Minister) to announce that no British assistance of any kind would be sent to the Poles who were trying to push back a Russian Bolshevik invasion. The reluctance of employers to encounter disruption while business was booming, and general war-weariness, aided the trade unions and the Labour Party.

The second cause for which trade unions were prepared to bring about widespread disruption was the coal mining industry, with its dramatic dangers, large union membership, strategic value, postwar price fluctuations, and declining labour productivity. Wage cuts were imposed by mine owners when a government subsidy ended in 1925 and they tried to go back on the 1919 agreement of a seven-hour day, just as miners were asking for a six-hour one and hoping for permanent government control or ownership of the mines. Wages and hours cuts were refused by the mining unions; their members were locked out by the employers. Conferences, committees, subcommittees and meetings with ministers failed to produce agreement, and the TUC called a general strike in 1926 to support the miners. Three and a half million workers were said to have responded.

However, it was not successful. The TUC's General Council, for reasons which are still a matter of argument, called off the strike on the ninth day, without a concession from employers or government. The miners remained on strike for a further ineffectual six months, returning on the brink of starvation. Disillusionment with the General Strike as a weapon, and with TUC leadership, subsequent restrictive laws (see section below) and rising unemployment have all been seen as the cause of a subsequent decline in strikes and in the appeal of a general strike. More peaceable relations between employers and employees followed for a few years, helped by talks between trade union leaders and employers, led by Sir Alfred Mond of ICI. Even before the General Strike, the collapse of the postwar boom had greatly weakened the shop steward movement, and a 'revolutionary' occupation of the Glasgow Town Hall in 1919 did not have universal support in the trade union movement. Trade union leaders from the late twenties concentrated more on the immediate interests of their members at work, concern for the growing numbers of unemployed, and resistance to changes which might decrease the wages or the number of jobs for their members. These leaders, most notably the able Ernest Bevin of the Transport and General Workers Union (TWGU) were vehement anti-communists. In spite of the disappointments of the General Strike and bitter recriminations over Ramsay MacDonald's decision to stay on as a 'National' prime minister when his Labour cabinet resigned in 1931, the trade union movement was committed to change through peaceable methods. Table 6.1 shows some of the changes in unions and in the economy which most affected industrial relations between the wars.

Trade Unions, Workers, and the Law

The Treasury Agreement of March 1915, made with trade union leaders by Lloyd George, was incorporated into the later Munitions of War Act. This banned all strikes and lockouts on war work, made arbitration compulsory and promised to end dilution when the war was over. In 'controlled establishments', those said by the Ministry to be essential to the war, all restrictive practices were banned, discipline was firm, and rule-breaking was dealt with by special Munitions Tribunals. The Act's most unpopular clause said that workers could not leave a munitions firm without a 'leaving certificate' issued by an employer. It was intended to prevent employers from poaching each other's labour and also to stop workers bidding up wages by going or threatening to go to better-paid jobs, or quitting munitions manufacture altogether by joining the armed forces. The excess profits duty was the sweetener to the unions. The introduction of dilution was delayed for about six months, however, because employers were unwilling to pay 'dilutees' full craft rates for their labour, and employees were unwilling to show them how to do the job.

Most wartime legislation and regulations ended with the war. A special Act, to ease the transition to peace, forbade the lowering of

Table 6.1 *Union Membership — General Indicators*

Year	*Total union membership (000s)*	*Total union density (per cent)*	*Number of stoppages beginning in year*	*Total working days lost by all stoppages in progress during year (000s)*	*Cost of living (1914=100)*	*Weekly wage rates (July 1914=100)*	*Percentage unemployed (yearly mean)*	
							of total workforce	*among insured workers*
1911	3,129	19.0	872	10,155	97	95		
1912	3,394	20.6	834	40,890	100	98		
1913	4,107	24.8	1,459	9,804	102	99		3.6
1914	4,117	24.7	972	9,878	100	100		4.2
1915	4,335	25.9	672	2,953	123	110–115		1.2
1916	4,611	27.4	532	2,446	146	120–125		0.6
1917	5,452	32.3	730	5,647	176	155–160		0.7
1918	6,461	38.1	1,165	5,875	203	195–200		0.8
1919	7,837	46.0	1,352	34,969	215	215–220		—
1920	8,253	48.2	1,607	26,568	249	270–280		3.9
1921	6,512	37.9	763	85,872	226	231		16.9
1922	5,573	32.1	576	19,850	183	191	10.8	14.3
1923	5,382	30.7	628	10,672	174	188	8.9	11.7
1924	5,463	30.9	710	8,424	175	194	7.9	10.3
1925	5,430	30.4	603	7,952	176	196	8.6	11.3
1926	5,152	28.6	323	162,233	172	195	9.6	12.5
1927	4,860	26.8	308	1,174	167½	196	7.4	9.7
1928	4,753	25.9	302	1,388	166	194	8.2	10.8
1929	4,804	26.0	431	8,287	164	193	8.0	10.4
1930	4,783	25.7	422	4,399	158	191	12.3	16.1
1931	4,569	24.3	420	6,983	147½	189	16.4	21.3
1932	4,395	23.3	389	6,488	144	185	17.0	22.1
1933	4,350	22.9	357	1,072	140	183	15.4	19.9

Source: Clegg, H.A.(1985) *A History of British Trade Unions Since 1889*, Vol. II, p. 568, The Clarendon Press.
Feinstein, C. (1972) *National Income, Expenditure and Output of the United Kingdom, 1855–1965*, Table 58 (this applies to column 8), Cambridge University Press.

wages for six months. The scope of the prewar Trade Boards which regulated wages and conditions in low-paid non-unionised workshops, such as East-End clothing sewing rooms, was extended. The Emergency Powers Act of 1920 gave the government power to govern by decree, as in a war, should the supply and distribution of food, water, light, or transport be threatened. It was used in the 1921 strike and the 1926 General Strike.

After the latter, the Trade Disputes and Trade Unions Act of 1927 was passed. It made all general strikes illegal as well as all sympathy strikes held by the courts to be intended to coerce the government directly or indirectly by imposing hardship on the general public. Government employees were to belong only to unions exclusively for state employees. Picketing involving intimidation was illegal, and the government was given a vaguely-worded right to seize union funds. 'Contracting in' replaced 'contracting out' for the political levy. This reduced the Labour Party's income by a third. The sympathetic strike clauses were never used, although they may have deterred some strikes. No other changes were made before 1939.

The Growth of Trade Union Membership

In 1913 there were just over four million trade unionists; by 1920 there were 8.25 million, but this was the peak interwar year for membership. During the war total industrial employment fell because of the millions recruited into the armed services, but union membership increased as numbers employed in metals, chemicals and government industrial establishments rose by about 1 million, and many of them (including women who went into these industries temporarily) joined trade unions for the first time. The Workers' Union, which recruited extensively among women war workers, became for a time the largest union in the country. The skilled or craft societies did not recruit women. As unemployment rose, and unions were defeated in strikes, membership fell to a trough of 4.4 million in 1933. Thereafter, as recovery got under way, it increased. By 1939 it had reached 6.3 million, still below the 1920 peak.

Postwar there was a spate of union mergers: the result of the feeling within the trade union movement, and particularly in its radical wing, that size represented strength, and that unions in particular industries should not find themselves in conflict with each other. The amalgamations were aided by the Trade Union (Amalgamation) Act of 1917. The reformers were not able to achieve their aims completely.

In manufacturing, the first important merger was in the iron and steel industry; others followed. 1922 saw the formation of the TGWU which absorbed small unions in almost every subsequent year, acquiring a significant membership in manufacturing when, in 1929, under the forceful leadership of Ernest Bevin, it took over the Workers' Union. By 1939, it had incorporated 45 unions. The union of General and Municipal Workers (GMWU) was formed in 1924.

Higher unemployment in the 1920s meant that wealthy (often skilled) unions were less disposed to merge with poor unions and were less keen to be in disputes which they might otherwise avoid. By 1939, with membership rising, unions competed with each other to recruit members and at that year's Trades Union Congress at Bridlington, the 'Bridlington Agreement' was concluded. It regulated recruitment and tried to prevent poaching of one union's members by another.

In World War I, in spite of strong local movements, many district-level collective bargaining structures were replaced by national-level collective bargaining or national level government arbitration which, in general, was retained. In 1917 the Whitley Committee proposed that every industry should have its own Joint Industrial Council (JIC) to determine wages and conditions nationally, in a spirit of industrial cooperation: the aim was to increase productivity, and to allow labour to take its share of the resultant increased prosperity. But many industries (especially well-organised ones) stuck to their own collective bargaining agreements; after 1921, few JICs were formed. By 1939 chemicals, cement, electric cables, flour milling, hosiery, paint and varnish, pottery, printing, silk, soap and candles, and tinplate had JICs. Their effectiveness varied widely.

Working Hours

In the war years the standard working week remained at 52–56 hours. In 1919 6.5 million workers achieved reductions (averaging 6½ hours per week) through collective bargaining without loss of pay. The average standard working week became 48 hours. The unsuccessful Glasgow 'revolution' of 1919 began with a strike to achieve a 40–hour week without loss of pay in order to spread available work. A local general strike was achieved, but not a 40–hour week.

Attempts were made (by employers) after 1920 to extend hours, usually unsuccessfully. An exception was the colliery owners who returned the miners from a seven- to an eight-hour day after 1926. In the 1920s only a tiny minority of manual workers had paid holidays. One week's paid holiday became standard for all workers whose conditions were governed by Trade Boards or Agricultural Wage Boards, and over half the workforce had paid annual holidays by 1939.

Wages

Inflation, which accelerated until 1920, was accompanied by rising money wages and after 1917 rising real wages (see Table 6.1). From 1921 until 1933/34 prices fell almost continuously and employers tried to reduce money wages, causing strikes. The government thought that one way to reduce unemployment was to cut wages, so that more employers could afford to take on more labour, but did not enforce this. Some reductions were due to prior agreement to relate wages to the cost of living or the price of the product. In the later 1930s, as recovery came, prices rose slowly, and so did wages.

For those in work, it is estimated that between 1913 and 1938 average real wages (that is, after taking account of inflation) increased by 30 per cent. Generally, slightly lower wages prevailed in areas of high unemployment, though national agreements kept up some wages in industries in such areas. Only with the rearmament of 1937 onwards did work and average wages pick up in Wales, Scotland and the North of England.

It has recently been argued that a good deal of the unemployment of the 1920s and 1930s was caused by workers voluntarily choosing to be on the dole rather than to work. This is unlikely to be true. The absolute level of dole payments as well as its differential from what could be earned must be taken into account. The low level of benefits did not provide a standard of living any worker could wish on himself or on his family, and the iniquities and indignities of the hated Means Test were no incentive to stay at home. There is no constant correlation in the years 1929–1939 between the real level of unemployment payments, under their successive names, and the degree of unemployment, nor between payments and real wages and unemployment, except the positive correlation between slightly lower wages in regions of high unemployment. Research is still proceeding on the effect of wage rigidity in old and new industries. It was the case that when benefits were made easier to claim, more people claimed them, however.

Conditions of Work

The less profitable and less prosperous a firm, the less attention it generally paid to welfare. Older factories and workshops, especially small ones, lacked elementary sanitation, and workers ate their lunchtime sandwiches amid the grease and grime of machinery. Thus there was a growing consensus that employers should upgrade the workplace, even if only for enlightened self-interest. ICI and Courtaulds, Rowntrees and Cadbury prided themselves on clean work surroundings, sports days and sports fields. Trade unions and work-people differed in the importance they attached to environment; many thought that money so spent should instead have been included in the pay packet.

Many wartime changes were transitory. A lasting one was in the increased proportion of women workers, mostly in assembly industries.

Another was in the proportions of skilled, semi-skilled and unskilled workers as production became a little more automated: in engineering these were 32, 57 and 11 per cent respectively in 1933. In 1914 they had been 60, 20 and 20 per cent. These figures should be treated with care, for they represent not only changes in the actual level of skills and the types of skill required, but also in the rather arbitrary classification of skills. Nevertheless, the move from skilled to less skilled workers is not just a matter of classification.

World War I stimulated the development of work-study by employers, since maximum efficiency of production required efficient utilisation of workers. This influenced developments in the interwar period. From 1920, for instance, the Process and Rate Department of one of the largest engineering employers, MetroVick, undertook time study, and in 1930 a special 'motion study' department was established.

A method of work control and payment known as the Bedaux System, after its founder, Charles Bedaux, but deriving essentially from Taylorist principles, gained some importance. It meant subdivision of work, close observation of work by specially trained managers, and a system of piece-rate payments. It was best suited to assembly industries such as radio manufacture. An investigation of its use by the TUC in 1933 led to reluctant acceptance, but individual unions thought otherwise. Strikes took place, but Bedaux remained. Lucas, after an anti-Bedaux strike, introduced a Bedaux-like system under the name 'Lucas Point Plant System', using Bedaux specialists, but with their identity carefully concealed. This reorganisation of work implied a decline in the role of foremen in controlling production, and an increasing importance of professional production engineers in organising it. Most unfortunately, it contributed to undervaluation of the worth and wider function of these professional engineers, for example in quality control, since their presence caused trouble, and most managers did not press their employment.

Engineering Unions

Nearly one third of all wartime strikes was in engineering and shipbuilding, whereas in 1899–1913 working days lost in these industries averaged only 16 per cent of the total.

The ASE was the most important of the engineering unions; its members, exclusively craft and male, increased from 161,055 to 298,782 from 1914 to 1918. A three-year payment agreement in the engineering industry on the Clyde ended in mid-1914. In December 1914, the ASE put in a claim for an extra 2d (less than 1p) an hour. The ensuing Clydeside dispute indicated the increasing rift which separated officials of trade unions from shop stewards. Postwar, the ASE official leadership acted firmly against its district committees and shop stewards in Glasgow, Belfast and London, keeping to the 47-hour agreement recently made with employers, and disowning support for the 40-hour movement. The employers themselves were hostile to shop stewards. In 1920 the ASE merged with a number of smaller craft societies to form the Amalgamated Engineering Union (AEU). Membership stood at a peak of 459,987, a figure not reached again until 1941.

In March 1921 the Engineering Employers Federation (EEF) rejected a pay claim put forward by the AEU, and instead demanded pay cuts on the 1919 increases because of the fall in orders and prices. The EEF was the least conciliatory and most hostile employers' organisation in manufacturing industry. It much resented government consultations with unions during the war. The AEU leadership and other union leaders refused to consider the cuts; the employers refused to go to the Industrial Court. The AEU after a ballot conceded the employers' pay demands, but continued to demand overtime only by agreement, control over the manning of machines, and the right to negotiate apprentices' payments. The EEF then demanded full control over 'Managerial Functions', most of which they were supposed to have under the 1898 Terms of Settlement. The executive of the AEU agreed, but a ballot of the membership voted against acceptance. The EEF then locked out AEU members from March 1922, and extended the lockout to cover all the other engineering unions. The dispute ended in June as a complete victory for the employers, at the cost of involvement of over 260,000 workers and 13,650,000 working days lost. The 1898 Terms of Settlement, theoretically suspended only for the war, were to be returned to, and wartime consultations abandoned. Even so, individual employers hung back from enforcing the Settlement's terms.

In 1929 the AEU, now including some semi-skilled men, had new demands but the employers, who had done rather better in 1928, would not grant even a wage increase. Although some of these demands were later achieved, the union was forced to accept cuts in rates in 1931 as prices fell, although it managed to safeguard the 47-hour week. Nevertheless, employers on the whole had still not introduced sweeping changes in the organisation of work.

More employment due to rearmament expanded the AEU's membership and strengthened its bargaining power after 1935. In 1937 it obtained paid holidays and, after a strike of apprentices, the right to negotiate for apprentices' pay. Some increase in wages was obtained, but not a 40-hour week. Relaxation of factory discipline followed the return to work of men who had not worked for many years; shop steward activity revived and increased. Twenty years of intermittent hostility had settled little within the union or between union and management.

Cotton Textile Unions

After a burst of activity when war began, shortage of raw cotton imports caused widespread unemployment. Workpeople left the industry. Imports then improved and an order had to be applied to stop more leaving. The Cotton Control Board (1917) had extensive powers to operate a rota system in the mills. It agreed that, when not working, workers should be paid through the unions from a levy imposed on the whole industry. The union leaders, however, did not play an important role in making policy, and wages did not rise as fast as the cost of living or the wages of other workers. Employers were organised in two strong federal organisations which helped them resist union pressure.

After an initial boom and wage increase, and a reduction of hours from 55½ to 48 in 1919, short-time working began in the winter of 1920. Cotton textile prices fell and employers, after a two-week lockout, cut wages, and did so again in 1922. From 1921–32 there were many strikes and lockouts, usually over pay; 32 million working days were lost in an industry which, apart from 1911–12, had been known for its orderly industrial relations.

In 1920 two cotton unions achieved their highest ever membership levels; there was much discussion about the establishment of one big textile workers' union; but the proposals foundered subsequently in the face of falling union membership. Unemployment in cotton mounted in 1928, ahead of other industries: wage cuts produced a lockout in 1929.

In 1929, to cut costs and preserve markets, employers and unions in Burnley began experiments in which each weaver (usually a woman) operated eight looms rather than the traditional four. By 1931 Burnley weavers, well aware of high unemployment in the industry, and irked by a small wage cut, decided to end the agreement. Burnley employers then locked them out and weaving employers throughout Lancashire followed suit. Despite the intervention of the Ministry of Labour, the lockout lasted four weeks and involved 113,203 trade unionists. The issue was left unresolved; some employers insisted on eight looms, others compromised on six. Others again — those with high-quality specialities — were content with four.

In 1932 some employers imposed a pay cut on four-loom weavers as orders fell. Another four-week strike followed, called off after painstaking negotiations by the Chief Conciliation Offices of the Ministry of Labour. The Midland Agreement (named after the Midland Hotel in Manchester where they met) was the outcome. It provided new collective bargaining machinery, with independent arbitration where necessary. The Weavers' Amalgamation then signed the More Looms Agreement which regularised the care of six looms per operative; yet four looms remained in much of the industry.

The Cotton Manufacturing (Temporary Provisions) Act of 1934 made wages in weaving determined by collective bargaining legally enforceable throughout the industry, a unique arrangement in the history of industrial relations. A more constructive atmosphere emerged, and from 1936–9 all the cotton unions, together with the employers, worked on plans to reduce domestic competition in the cotton industry.

Shipbuilding Unions

Shipbuilding retained its rigid separation between craft skills. Shipyards employed men whose skills were specific to shipbuilding as well as others — boilermakers, plumbers and electricians — with less specific skills. From 1914 to 1921 shipyards were busier than ever and employment and trade unionism, already strong, expanded. Hours were reduced to 47 from January 1919. The war and the radicalisation of trade union thought led, as in many other industries, to an impetus to merge unions to increase strength and avoid disputes between themselves, but with heavy unemployment in the industry after 1921, rivalries broke out and the merger did not take place.

As prices and orders for ships fell in 1922, employers pushed for wage cuts of about 25 per cent on their peak money level and this led to disputes. The shipbuilders were locked out for six weeks before agreeing. In 1923, too, there was a major dispute. The unions wanted to limit the amount of overtime worked to increase employment, but employers considered the amount of overtime to be their responsibility. In 1923 agreement was reached essentially on the employers' terms. However, the Boilermakers' Society refused to accept it and were locked out between April and November 1923, leaving the industry paralysed. Finally, a compromise based on the initial agreement was reached.

The boilermakers were not pleased, and in 1924 left the Federation of Shipbuilding Trades. Other crafts — electricians, painters, patternmakers, shipwrights and woodworkers — did likewise. Only a few minor crafts and the general unions stayed in. The rest, except for the boilermakers, formed a Shipbuilding Trades Committee. This committee, and the federation, made a new procedural agreement with the employers, but the boilermakers twice voted by ballot against it. The employers in 1929 made offers on piecework to tempt the boilermakers. Twice more they were rejected; nevertheless, pay negotiations continued with all the relevant unions. The employers were not inflexible. Union divisiveness remained; a small cut in money wages for both piece and time workers in 1931 was accepted under protest: they were at least gaining nationally uniform wages. This made subsequent negotiations easier, though it made any attempt to relate the wage bill of a particular yard to its ability to pay more difficult.

For the remaining interwar years there were very few strikes. Unemployment of over 60 per cent dampened protests. The 1930s saw many shipyards dismantled. Workers from Jarrow marched to London in 1936, making their town a symbol of the unemployment and misery of the 1930s. Inter-union disputes and management–union arguments over new techniques in shipbuilding were not resolved. The title Confederation of Shipbuilding and Engineering Unions (CSEU) replaced the old Federation in 1936, but it was still a loose and not very comprehensive grouping.

Iron and Steel Unions

In 1915, the Iron and Steel Trades Confederation was formed. Because the 1871 Trade Union Act made mergers between unions very difficult until altered in 1917, an ingenious scheme was devised by which a new confederation, the Iron and Steel Trades Confederation (ISTC), and a new union, the British Iron, Steel and Kindred Trades Association, would be formed. The ISTC would undertake all negotiations. Some unions at first failed to obtain majorities for this in ballots, but joined in 1921. The Blastfurnacemen's Federation (later the National Union of Blastfurnacemen) refused to join. The ISTC began life on 1 January 1917 and its first president, John Hodge, was a Liberal MP and Lloyd George's Minister of Labour. Wages and earnings increased during the war.

In most industries, postwar reduction of the working week to 47 or 48 hours was a drop of three or four or at most nine hours. In iron and steel, a 70- or 72-hour week with 12-hour shifts was the norm except in South Wales and Scotland in 1914. To reduce hours to 48, a third shift had to be worked to make the best use of the furnaces and to process smelted metal. It was unlikely that productivity increases could cover the cost to employers of a third set of wages. The ISTC in 1919 agreed that, if existing wages were kept for the lower-paid, largely time-paid workers, higher-paid furnace workers, some earning up to £11.50 weekly (four times the earnings of low-paid men) would take a cut of up to a third in their income when hours were reduced.

When output and prices of steel rose in 1920, smelters and other senior men with wages related to tonnage produced received more and faster increases. Time workers wanted to benefit too, as they did in the war, and it was agreed by the employers, the ISTC, and the National Amalgamated Union of Labour (which had been recruiting unskilled men ignored by the ISTC) that their wages would be adjusted to selling prices. When prices fell in 1922 some timeworkers, particularly the bricklayers who saw to furnace lining, claimed that they had not subscribed to the agreements. The Building Trades Union made a compromise agreement for them.

By 1923, while there was widespread variation, average money wages were just over half, and in real terms just over three quarters, of the 1920 peak. To help lower-paid workers, the ISTC agreed to reinstate weekend work with its overtime pay, but a more thorough-going rethink about pay structure was needed. There were more and more men in iron and steel who were neither labourers nor readily identifiable craftsmen, but who undoubtedly possessed skills as, for example, crane drivers or operators of charging machines for furnaces. The rethink was postponed. Profit margins in steel, if they existed at all, were low, and foreign competition was evident. The ISTC agreed that the smelters would, in 1928-29, take an average cut of 13 per cent to give an increase to the lowest paid. The smelter's income, in money terms, had risen to its 1919 level (and was higher still in real terms) because larger furnaces increased output. It was also agreed that 'tonnage' rates would be attached to all work which could be related to output. This certainly gave a general incentive — and once more by-passed branch organisations whose leaders were more militant than the ISTC's national leaders.

The General Strike of 1926 received only guarded support from the ISTC, even though its secretary was president of the TUC's General Council. In the worst years of the Depression, the steel industry had high levels of unemployment (46.2 per cent in 1932). After 1932, unemployment decreased; improvements in wages, the introduction of holidays with pay, and in one large firm, United Steel, the establishment of a pension scheme for workers, followed. It was an industry which showed some flexibility and goodwill in industrial relations.

Chemical Unions

In the chemical industry effective collective bargaining was established during the war. Wages and conditions could have been set by the Committee on Production. The employers, however, wanted to avoid the delays this procedure entailed and approached the trade unions. In 1918 the Chemical and Allied Employers' Federation and the Heavy Chemicals Joint Industrial Council were formed. Employers granted the 47-hour week in 1919; after 1921 money wages were cut, but comparatively high wages in real terms prevailed.

ICI's wages and conditions set the pace for the rest of the industry. On its formation it had 33,000 employees and quickly developed a coherent and centralised labour policy which owed much to the policies of Brunner Mond. Sir Alfred Mond took a great interest in labour matters within industry as a whole. However, his own policy at ICI had as one of its principal aims the discouragement of traditional trade union activity, in any case weak within the chemical industry. Within ICI, union membership, a low 30–40 per cent, decreased during the Depression as employees were temporarily dismissed.

The company took active steps to bind the workers 'more closely to the Company's interests than to the interests of the "Working Classes"' (Reader, W., 1975, *A History of ICI*, Vol. 2, p. 64, Oxford University Press), paying good wages by the standards of the day. Works Councils of managers and elected workers' representatives met monthly. Above them were General Works Councils covering a number of factories, and above these was the Central Works Council, presided over by the chairman of ICI. The Councils were purely consultative, however, and barred from discussing the most contentious issues in industrial relations — wages, hours and conditions of work.

Employees could buy ICI shares below market price, and the Staff Grade Scheme allowed management to promote to the Staff Grade hourly paid workers who had worked for the company for more than five years. This gave important advantages — guaranteed weekly wages, one month's notice of discharge, paid Bank Holidays, and full wages during sickness. However, only 25 per cent of those eligible were given staff status and the unions disliked the scheme intensely because of the power it gave management. Only later did management recognise the resentment the scheme caused.

In 1935 ICI left the Chemical and Allied Employers' Federation. With a generally expanding market, it invested capital, enlisted worker support in increasing productivity, and gave employees a share of increased profits. It had a coherent and controlled approach to labour relations, though the unions did not entirely like it.

Rayon Unions

Rayon in 1935 employed nearly 35,000 workers. Output had increased faster than employment because of substantial increases in productivity due to technical improvement of processes and an increased pace of work. Union membership was generally low and the two unions with the biggest membership were general unions, the TGWU and NUGMW. Courtaulds, like ICI, was once known for its anti-trade union sentiments; it took longer to develop a unified and coherent labour policy. There were labour troubles due to poor working conditions (an initial feature of the industry which produced smelly waste, though substantial improvements were made) and the attitudes of some managers. Further trouble in the early 1930s led to Courtaulds' introduction in 1937 of a centralised system of industrial relations and to a week's paid holiday. As part of this new order the TGWU was recognised, and many issues were dealt with directly between senior managers and Transport House, headquarters of the TGWU. No matter how enlightened a firm's management might be, it could not kill trade unionism by kindness in the Britain of the 1930s.

Motor Vehicle Unions

Employers in the motor industry were almost universally actively hostile to trade unionism, and most avoided recognition of it and of shop stewards until the 1939–45 war. Most of the large firms were 'non-federated', that is, they did not belong to an employers' federation, in this case the EEF, and thus were not subject to the agreements made between the EEF and the engineering unions. The motor industry's mass production methods had important consequences for the organisation of work and payment systems, though each company had its own system. Semi-skilled workers came to predominate in production, with skilled workers employed only for specialist work in the toolroom.

The British Ford Company was controlled from the head office of the American parent in Detroit. Day-work payment combined with strict supervision of work and output and a strong antipathy to trade unions was Henry Ford's US policy, applied in Britain first at Trafford Park in Manchester (though visitors from Ford USA thought this plant slackly run) and then at Dagenham. This was in sharp contrast to the engineering industry as a whole, where piece-rate payments were increasingly important. Ford dealt severely with any emergence of trade union activity: in 1933 a strike at Dagenham against wage cuts led to the sacking of its leaders. In the late 1930s there was an unsuccessful campaign to unionise Ford in which the TUC itself was involved.

The other American firm, Vauxhall, was more conciliatory to trade unions. From 1928 a 'Group Bonus' method of payment was instituted, requiring less supervision than the Ford system and this, together with a profit sharing scheme (from 1935), relative stability of employment and the relatively high level of wages, reduced the attractiveness of trade unions, which made very little headway in recruiting members.

The British-owned Austin and Morris were also hostile to trade unions. Morris kept them out, paying high wages and cultivating good relations with employees through sports days, brass bands, and so on. Sudden lay-offs when sales fell undermined some of the goodwill engendered, however. Morris used as much automation as practicable, but relied on group and individual piece rates to stimulate output. In its Oxford works, trade unionism was resisted despite a strike in 1934 at the neighbouring Pressed Steel plant which produced car bodies for a number of firms including his, and which led to the recognition of unions there. At Austin's works in Longbridge there were unsuccessful attempts to establish shop steward organisation. There was, however, a major strike at Longbridge in 1928 over wages and the piece-rate system. Austin was a member of the EEF and therefore agreed to negotiate with trade unions, the only major car producer to do so in this period.

In the motor industry, therefore, trade unionism had little impact, though determined men tried to establish it. The level of unemployment in the economy as a whole and the seasonal character of much of the employment in the industry played a part in limiting unionism but did nothing to lessen antagonism.

Conclusion

Trade union and worker cooperation was essential to the government in wartime and so strikes and threats of strikes had considerable impact. Union leaders were consulted on a variety of issues although the Clydeside claims for workers' control of industry proved fruitless. In boom conditions after the war trade unions tried to exert the same kind of pressure. They were successful over the Russian and Polish issues, but not over coal mining. In peacetime, government priorities were different, and after 1921 pressure was not as effective, partly because employers were less dominated by the need to meet orders.

Trade unions which merged in this period are among the largest and most powerful still. The importance of national collective bargaining emerged. Finally, the period between the wars was important because of the attitudes brought about by unemployment and money (and sometimes real) wage cuts.

Further Reading

Charles, R. (1973) *The Development of Industrial Relations in Britain*, Hutchinson.

Clegg, H.A. (1985) *A History of British Trade Unions Since 1889*, Vol. II, The Clarendon Press.

Hinton, J. (1973) *The First Shop Stewards Movement*, Allen and Unwin.

Pagnamenta, P. and Overy, R. (1984) *All Our Working Lives*, British Broadcasting Corporation.

Specific Occupational Groups

Burgess, K. (1986) 'Authority relations and the division of labour in British industry with special reference to Clydeside', *Social History*, Vol. II, No. 2, May, pp. 211–233.

Questions

(1) How successful were strikes, or the threat of them, in getting unions what they wanted, 1914–1919?
(2) Why were demarcation disputes important in the shipyards, and what was done to avoid them between the wars?
(3) How and to what extent did the changing fortunes of *two* of the industries discussed in this chapter affect trade unions in them?
(4) Why did the 'density' of trade union membership and the structure of unions change in this period?
(5) What did the contest between the engineering unions and the EEF achieve, 1918–1939?

Section C

1939–1951

World War II: [illegible] of the Supermarine Spitfire

7

British Manufacturers and the Economic Environment

This section covers World War II (1939–45) and six years of postwar Labour government. 1939–51 was an epoch of unprecedented direction and control by government of British manufacturing and the British economy. The economy still had a high degree of dependence on imports and exports, and so the wide spread of the war and the major economic featues of the postwar world will be outlined. But first a number of trends which made government intervention more acceptable in both war and peace will be considered.

Perspective: Plans and Practicalities

Hitler's rise to power in Germany was viewed with popular distaste and alarm in Britain in the mid-1930s. While politicians might bargain for a temporary delay, there was a feeling that sooner or later he would have to be stopped. Within government circles there was acknowledgement that if war came, it would be likely to last at least three years and would demand full mobilisation of Britain's resources. Germany took over Austria and Czechoslovakia unchallenged, but when Hitler threatened Poland, France and Britain were committed to war against him. When France was defeated in 1940, and Britain stood alone in Europe, there was no doubt that the British economy would be directed with one aim — maximum war production — and only minimum attention to civilian needs.

The idea of a directed or planned economy in peacetime had also been gaining ground. The Political and Economic Planning group (PEP) wanted to make better apportionment of investment, to promote efficiency and economic growth, within a capitalist framework. Socialists such as G.D.H. Cole, who wrote *Practical Economics*, a thought-provoking book, were more concerned with government *ownership* and direction of the economy to end unemployment. Labour politicians wrote about planning and nationalisation. These ideas, together with the practical achievements during World War II, prepared many people for more government intervention after the war.

In 1930, the Labour Government appointed an Economic Advisory Council. Its influence was brief, but Sir Josiah Stamp, one of its two economists, began an economic and financial survey of Britain along the lines of surveys in the USA and Germany. His aim, however, was to provide a tool for a possible war economy, and indeed it was the basis for wartime planning. The Council was reshaped as the Central Economics Information Service. Later divided into two, its information became more detailed, and its recommendations and information were used in war by the Cabinet. An inadequate system of giving 'priorities' to different ministries and authorities was replaced by broad allocations to them. The allocations roughly added up to the sum total of resources (including manpower) available.

Conservatives were unlikely to accept such a profusion of controls in peacetime, but they did accept that there should be no return to prewar unemployment, eliminated during the war. Therefore, when economists recommended demand management, they agreed — not least because it seemed to promise a reward for the efforts made during the war, and because economists, who had been essential to war planning, had apparently learnt much more about the workings of the economy. Labour leaders accepted it too, but in their victorious 1945 election campaign also promised nationalisation of key industries, and a government-financed housebuilding programme to replace damaged dwellings and to compensate for the virtual end of housebuilding for seven years.

However, the big postwar economic problem was not unemployment, but the need to increase exports to even out the balance of payments and to retain the value of sterling both to keep faith with creditors who held sterling, and to prevent sterling leaving the country in anticipation of a devaluation which that very outflow would make more likely. All political parties agreed that some controls were necessary to deal with these problems. The Labour Government for a time continued wartime allocations to manufacturers and continued rationing to consumers to promote exports and yet provide 'fair shares' to a war-weary population. Under the circumstances, Conservatives accepted that much of this was necessary, and so criticised the manner in which things were done rather than that they should be done at all.

The Spread of the Conflict

In 1940 Germany invaded Denmark, Norway, France, Holland and Belgium. Each surrendered. Italy joined Germany, and only the Royal Air Force and the Channel saved Britain. Russia, which had a pact with Germany, had taken part of Poland, and the Baltic states. Germany then turned on Russia which thus became Britain's ally. Japan wanted oil and rubber for its long war in China and in 1941, to gain supplies in the Pacific, attacked the US fleet in Pearl Harbour. The USA then became an ally and, seeing Britain's isolation and the terrible slaughter in Russia, promised to defeat Germany before Japan. This was done, with Russia reviving and turning Germany out of Eastern Europe, and civil wars took place in the Balkans. Fighting stretched from the South Pacific to the Arctic Circle. Bombing did damage, but much worse damage resulted from retreating armies destroying buildings, bridges and food supplies. Russia reckoned that 20 million of its people died because of the war; 6½ million people, most of them Jews, died in Hitler's concentration camps. Over 50 million people died altogether. Millions of 'displaced persons' were in hutted camps in Europe, unwilling or unable to return home. The allies had proclaimed freedom from want as a war aim: achieving it seemed impossible in 1945.

Features of the Peace

During the war the US economy expanded rapidly. It offered 'Lend-Lease' of defence equipment (very widely defined) to Britain, and some to Russia. When the war against Japan ended abruptly after two atomic bombs were dropped, Lend-Lease ended too, but the USA cancelled all obligation to return or pay for the material. It then gave money to damaged countries through the United Nations Relief and Rehabilitation Administration, then offered loans (though the conditions attached to them were hard ones), and next, after the severe winter of 1946/7, aid for countries wanting to rebuild their economies. This was first suggested by the US Secretary of State, General Marshall. The USA also fostered the World Bank, the International Monetary Fund, and GATT, the General Agreement on Trade and Tariffs. There is no doubt that the USA was seeking to rebuild the world in a capitalist image, but a less restrictive image than the prewar tariff-and-subsidy one. US involvement in world affairs was in marked contrast to its isolationism in the 1920s. Even the money the US spent on its troops abroad was an asset to weakened economies.

In part because of US expenditure, countries began to recover, though Russia and the countries it had occupied (except Austria) were increasingly shut off from the rest of the world, as political attitudes hardened. The Korean War of 1950 threatened to upset the political and economic balance being achieved (Russia aided North Korea, US forces formed the major part of the United Nations troops sent to South Korea), but in spite of doubling and even tripling of the prices of commodities ranging from wool for winter uniforms to copper for electrical insulation, there was no lasting check to growth. Germany and Japan began to rebuild and replace their destroyed and damaged industries, and neither, postwar, spent much on the armed forces.

Complaints, Criticisms, and Controversies

Most arguments about economic development in Britain, 1939–51, turn on the wisdom or otherwise of government policies. There is agreement that opportunities were missed before May 1940 when Churchill became Prime Minister. Some erstwhile planners and controllers, postwar, wrote scathing criticisms of the methods used 1940–45, yet did not suggest alternatives.

Stronger criticisms are made against controls and plans, 1945–51. Instead of government intervention, would it have been better to let prices, the balance of payments, even the value of sterling, take care of themselves? Would unrestrained domestic demand have unleashed a new surge of British production? Was the Labour Government's ideal of a more equal society with the rationing and controls it involved a barrier, psychological if not physical, to greater growth? Or should the inadequacies of British management have received selective government intervention? Did nationalisation have the benefits claimed for it? Some of the analysis needed to answer these questions is outside the scope of this book but, as in previous sections, we shall offer some points for consideration after we have considered individual industries in the next chapter.

Features and Factors

Trend figures for output and investment, for example, for the war years alone, assembled on a basis comparable with earlier figures in this book, are not available. Figures exist for 1937–51, and of course they average out rises and falls. From 1937–51, manufacturing output grew 2.5 per cent per annum, labour input 1.0 per cent and capital input 2.9 per cent; this means that total factor productivity in manufacturing grew at 0.9 per cent per annum — low compared with the periods which preceded and followed it. Nevertheless, in absolute terms, manufacturing contributed more to the growth of the economy than any sector except commerce (banking, insurance etc.) which, postwar, was making up for lost time. While other economies were still struggling, the British economy, measured by its very high employment levels, was at full stretch, yet much dependent on dollar gifts, loans, and earnings.

Size of Firms

1939–51 saw an increase in the size of plants owned by firms, rather than of firms themselves. In 1935, the percentage of total employment in plants employing 1,500 or more workers stood at 15 per cent. By 1951, this figure had risen to 24 per cent (Hannah, L., 1976, *The Rise of the Corporate Economy*, p. 159, Johns Hopkins University). There were a number of reasons for this. Firstly, some of the merged firms of the 1920s and early 1930s took time to build larger new plants to replace old ones, and these came into use only during the war. Secondly, government 'shadow' factories, conceived from the outset on a large scale, were built from 1937 onwards to be managed by existing firms for war production. Postwar, most of them were bought by the firms which had been using them. Thirdly, economies of scale were recognised even in medium-sized firms, and they extended their premises. The trend towards larger plants was particularly evident in aircraft, building materials, chemicals, engineering, and metal manufacture, although occasionally large firms built additional medium-sized plants, rather than big replacement ones. There was an assumption, especially postwar, that big, if not beautiful, was at least efficient, and government sponsored visits by manufacturers to the USA to look at large-scale production. The extent of rebuilding, re-equipment and capital expenditure necessary to imitate US production methods seems to have frightened off British producers, used as they were to a shortage of raw materials (especially steel), building delays, and difficulties in introducing new working methods.

At the other end of the size scale, the numbers employed in small plants of 200 or less fell from 44 per cent of the manufacturing workforce in 1935 to 35 per cent in 1951.

Large firms did not increase in stock market value as much as middling ones, and this reflected their rather lower percentage of profits. These were sometimes due to overseas wartime losses and to the absence of leading men from large firms — they were absorbed into governmental committees on rationing, allocation, and so forth. Further, these men may well have felt that their firms had a moral obligation to set a good example in observing controls and curbing profits. Smallest firms were sometimes closed altogether during the war, and production compulsorily concentrated in medium to large firms called nuclei. Allocations of raw materials on the basis of previous use made it difficult for new firms to get started. Big firms were thought of as the prime suppliers of goods — and jobs.

Mergers and Concentration

Merger activities slackened, 1939–51, in part because the largest firms had lower profits than pre-war and therefore less money to spend on acquiring other firms. Also, the government confined capital issues, that is, new shares issued through the Stock Exchange, to essential purposes. Further, because governments controlled prices and, to a degree, output, and manufacturers had strong agreements and understandings between themselves which curbed competition, there was less incentive to buy up a possible rival. Competition could be circumvented in other ways. Again, some of the earlier mergers had shown the difficulties of managing large empires and this was not encouraging. Indeed, in the paper and publishing industry postwar, some sections were 'de-merged' — sold off as separate enterprises.

The Labour Government 1945–51 was anxious that prewar price-fixing among firms should not reappear when government controls eased. By the Monopolies and Restrictive Practices (Inquiry and Control) Act, it set up the Monopolies Commission in 1948. When the Board of Trade requested it, the Commission could enquire into and make recommendations about activity which might not be in the public interest, by any firm or close group of firms with 25 per cent of the market share of any given product. However, while its reports gave ample evidence of collusive price agreements, its recommendations did comparatively little to modify them — let alone outlaw them.

Concentration increased overall in these years because of the faster growth of middling firms and the falling away of small ones, in spite of a decline in concentration at the top of the size scale of manufacturing industries.

Location of Industry

The government's first concern during the war was to put new industrial development out of range of German bombers. Lord Beaverbrook, Minister of Aircraft Production, tried a policy of 'dispersal' for aircraft factories, but this was inefficient. Even after the bombing of Birmingham and Coventry, while some factories were moved, others were simply patched up or rehoused in requisitioned premises nearby, to stay near their workforce.

Postwar, the government, fearful of a return of regional unemployment, continued prewar regional help. The Distribution of Industry Act (1945) increased the boundaries of Special Areas, renaming them Development Areas. All prewar help except abatement of taxation was available. The first munitions factories to be taken out of production and offered for sale were in those Areas. Further, Industrial Development Certificates, required by any firm opening new premises, were readily granted in the Areas, but reluctantly elsewhere. Finally, the Town and Country Planning Act (1947) gave local authorities power to prohibit unwanted development, and this helped the Areas indirectly. From 1945–7, two thirds of new employment and more than half of new starts in manufacturing were in the Areas, although they contained only 20 per cent of the population. After 1948, as government expenditure under the 1945 Act tapered off, export industries were allowed sites in regions where they most wanted to be.

The policy, at a time of high employment, was a success. Housing in the Areas was almost undamaged; it made economic sense to have industry where services and housing were intact. Unemployment in the Areas, postwar, was *lower* than unemployment had been in the most favoured pre-war localities.

Population, the Manufacturing Workforce, Employment, and Education

From 1931 to 1951, the population of Britain grew slowly from 44.8 million to 48.9 million. The proportion of elderly increased and the birth rate fell still further, reaching an all-time low between 1940 and 1942. On the other hand, partly because of increased attention to infant welfare during and after the war, infant deaths fell from 64 per 1,000 in 1930–2 to 29 per 1,000 in 1950–2.

Five million active people were in the armed forces by 1945. Some were performing services — cooks, quartermasters, the pay corps — which would have been needed were they in civil life. But most were not in productive employment. To supply them with war material, over 3 million people were in munitions manufacture. The participation rate of the civilian population rose. More married women worked; retired people did light assembly work in their own homes. Dilution was again agreed. Unemployment, still over 1 million in 1940, sank to a record low of 100,000 in 1943. Postwar, there was a decline in the number of married women working (perhaps related to the slight increase of family size). The increase of the school-leaving age to 15 in 1947, educational opportunities for ex-servicemen and women and two years' 'national service', usually in the armed forces, for all fit young men after 1950 removed people from the job market when demand was high. Unemployment at 1.5 per cent was a record low, postwar.

The workforce received a little more education and skill training. In the forces, there was much technical instruction and a good deal of general education through discussion groups and lectures on current affairs. On the one hand, the value of applied science for defence production was appreciated; on the other, a reaction from the scientifically-based weapons and horrors of war led many to devalue applied science in peace, though government and industry appreciated its importance more than before.

In addition to enlarged grammar schools, to which about 30 per cent of children would go, the Education Act of 1944 provided for secondary modern schools intended to cater for young people's all-round interests, and technical schools having an element of vocational training. Given the history of British attitudes, it is not surprising that this third type of school, excellent as many were, existed only here and there.

Capital

War had become more costly than in 1914–18. Urged on by Keynes, the government tried to meet as much war expenditure as possible by taxation; nevertheless, it borrowed two thirds of it, some by overseas loans, some by British loans. In order to avoid the high debt charges which followed World War I, interest rates were kept down. Bank rate fell to 2 per cent. There were tight controls on expenditure and so money went into government loans. *Banks*, by a gentlemen's agreement, restricted loans to favour defence industries, exports, coal mining, and agriculture. Capital expenditure on fixed equipment and rolling stock was postponed whenever possible; private investment virtually ceased. Shortage of capital did not hold up war production but only US help kept it going. 28 per cent of Britain's total net assets were lost during the war. New *stock exchange* issues were closely controlled.

Postwar, interest rates were still kept down, and capital issues and bank loans still restricted. Low interest rates were not only to assist central government and its war debts, but to help local authorities with their expenditure, particularly on housing. From 1939 to 1951, capital input into manufacturing industry rose 2.9 per cent, or 1.9 per cent per man-hour — not a big increase, but higher than prewar. By far the greater part of it came from ploughed-back *profits*. Industries which borrowed from banks benefited from low interest rates and preferred loans to raising money on the stock exchange. The interest rate was an encouragement to borrowing, but controls and allocations limited investment, and the shortage of dollars limited purchases of steel and equipment from the USA.

Export of Capital

Capital exports were forbidden as soon as war began except for some war-related purposes. Total British investment abroad in 1939 was about £4,000 million. One third of this was sold off to pay for immediate imports, and by 1945, income from overseas assets was less than one third of its 1938 money value. Postwar, controls continued except for a short period of relaxation under the terms of a United States loan. Then, with sterling convertible into dollars, capital went abroad until it was curbed. Lower returns from the smaller overseas investments did not help the balance of payments. A government endeavour to use capital exports to promote the growth of groundnuts in Nigeria and Tanganyika to benefit both these countries and Britain was a complete financial and ecological failure. Net investment income from abroad, which paid for about 25 per cent of British imports in 1936/8, covered only 15 per cent in 1948/50.

The Domestic Market

From September 1939 to May 1940, demand grew as ministries placed orders, and previously unemployed people spent a wage often at least double their previous 'benefits'. Afraid of a rise in the cost of living index which might spark off wage claims, the government subsidised foodstuffs — well-represented on it — and taxed items (including alcohol and tobacco) outside it more heavily. By 1941, price control, rationing of clothing and essential foodstuffs, and allocations were in full swing and demand was suppressed and diverted into savings. War goods, made to government order, were purchased as soon as they were produced. Blacked-out lights and sand-bagged buildings (to limit the effect of bomb blast) created an atmosphere of 'making do' and 'can't grumble'. Planners devised a 'points' system for clothing coupons and a similar one for non-essential (usually tinned) foodstuffs, keeping supply and demand in balance through non-money controls. The government permitted controlled prices to rise only if unavoidable costs rose. Overall, prices and wages increased by about 50 per cent, 1939–45: half the 1914–1919 rate. There was an illegal 'black market' in little luxuries.

Postwar, many controls, allocations, subsidies and rationing continued and demand did not push up prices as it might have done. In 1945, half of a typical consumer's expenditure was on price-controlled goods. Supply to the home market was restricted by the government because it wanted exports to earn dollars and other currencies to pay for necessary foodstuffs (including grain) and oil. Matters improved as manpower became available with demobilisation, munitions factories closed, and international trade revived. Then the Korean War and the decision to build up defence meant the reintroduction of some allocations of raw materials, and the retention of some food rationing. So once more demand was held down. 'Utility' standards for clothing and furniture, introduced in 1942 to save materials, were retained. Some utility items were appealingly simple and reasonably sturdy, even elegant. By 1951, there was a greater diversity of goods, but it was still very much a seller's market. Prices rose only an average 5 per cent per annum, 1945–9, but almost 20 per cent in the next two years because of devaluation and the Korean War.

Export Markets

During World War II, manufacturers were allocated extra materials or ingredients if producing exports needed to earn essential imports. Markets were easy. Postwar, this continued. Manufacturers could charge whatever they thought the export market would bear, and usually made much higher profits on exports than on domestic sales. There was much goodwill towards British products in the Commonwealth and the USA because of admiration for the British war effort. In 1945, the volume of exports was less than half that of 1938: by 1946 it was almost the same, and by 1950, 75 per cent higher, a remarkable feat. There was less import content in exported goods, partly because the export of cotton goods declined while that of engineering products, based on home-produced coal and mostly home-produced iron ore and steel, rose. But the terms of trade — the quantity of imports which could be purchased by a given quantity of exports — were not nearly as favourable to Britain as in the 1920s and 1930s. Britain, however, regained a sizeable share of world trade in manufacture, rising from 21.3 per cent in 1937 to 25.3 per cent by 1950. Defeated Germany and Japan had faster-growing economies, but started from a lower base in 1945. The USA was now both the major customer and the major competitor, though many of its goods were too expensive for postwar Europe.

Balance of Payments

In wartime, loans from abroad and sale of overseas assets helped bridge the gap in the British balance of payments, and postwar, exports were fostered to do the job. Until they could be revived, a loan was negotiated from the USA and Canada, effective in mid-1946. The declining terms of trade and a bitter winter which created a fuel crisis and held up exports meant that the loan was used up far more rapidly than intended. Further, as a condition of the dollar loan, currency controls had to be removed so that sterling could be freely converted into dollars. This had to end when the balance of payments went into deficit, and severe currency and import controls were imposed. After that, and aided by the Marshall Plan, and in 1949 by a devaluation of sterling in terms of the dollar, the balance was a little steadier, though a source of anxiety, until 1951. The government was constantly aware of the importance of exports of manufactured goods in keeping the balance right, compensating for the loss of overseas income.

The Role of Sterling

Sterling was not freely convertible to other currencies during the war; friendly sterling holders agreed to leave balances in London, although they knew that Britain's reserves had been run down. Partly because of their help, sterling was the most sought-after European currency postwar: its value in dollars held at £1=$4.03. Government controls and growing exports helped this. Then in 1949, a slight recession in the USA triggered off a decline in British exports. Fear that this might mean devaluation led to speculation in sterling: there were rumours that the US government thought the £ overvalued, and this led to more uncertainty and some withdrawals. The Labour Government, reluctant to devalue since, amongst other things, it might give it the reputation for being 'unsound', and would devalue the friendly sterling balances, postponed it, and then devalued by 30 per cent, making £1=$2.80. This was probably more than necessary, but the drop was intended to make it 'once for all'. Almost all the major currencies except the dollar followed suit. Consequently it was no longer as profitable to export to the USA, since less sterling was earned. Sales should have been easier but it was difficult to increase the volume of exports to compensate for lower dollar earnings. Then price rises of imports, due to the Korean War, produced new rumours about sterling. However, these were dealt with by other means.

Sterling clearly was not strong enough to withstand rumours and speculation unaided. In the interest of trade, and of invisible earnings, any government in the uncertain postwar years was bound to impose controls to sustain its value.

Government and the Economy

Great powers were given to the government between 1939 and 1945 under the Emergency Powers Act. All imports and exports, and the purchase and sale of almost all foodstuffs and raw materials, were in government hands. Exports, especially to earn dollars, had to be allowed for until Lend-Lease made dollars less necessary.

Government incentives to improve working conditions to keep up workforce morale included allocations of unrationed food to newly-expanded works canteens. Acceptance by both major political parties of Beveridge's ideas for an extended insurance scheme signalled a postwar extension of payments by employer, employee, and government to cover a wider range of benefits which would be available to all through the National Insurance Scheme of right, not through means-testing. Extension into peace of controls meant an expansion of the responsibilties and expense of peacetime government. Manufacturing industry cooperated — for instance in deducting Pay-As-You-Earn income tax from paypackets. Unexpectedly, demand-management was not needed to stimulate domestic demand, but to damp it down — both to encourage manufacturers to export, and to limit price rises. The government did this by increasing taxation, thus reducing spending power, and by cutting back its own investment and that of the private sector through its allocation of materials, building permits, and so on.

Two new ventures for government did *not* have all-party support. One was the creation of a National Health Service. This does not concern us here. The other was the beginnings of nationalisation in 1946. Labour's 1945 election pamphlets stressed economies, efficiency, safety, lower charges, and the prevention of competitive waste through nationalisation, that is, putting an industry into public ownership. Only one *manufacturing* industry was nationalised before 1951, and that was iron and steel. The nationalisation Acts of the Bank of England, Cable and Wireless, Civil Aviation, Coal, Electricity Supply, Gas Supply, and Transport (railways and road haulage) stayed on the statute books when the Conservatives returned to power (1951). Nationalisation certainly provided the opportunity for economies of scale and coherent planning within industries.

The fuel crisis of 1946/7, caused by an inadequate supply of coal and of rail rolling stock to move it to power stations, and made worse by weeks of freezing conditions in which railway points stuck and coal was frozen into heaps, was a test of the government's ability to predict and control crises in the nationalised coal and soon-to-be-nationalised electricity supply industries, and of its priorities in the economy. Predictions about weather and coal supplies were wrong; eventually it closed much of manufacturing industry rather than risk power supplies to hospitals, sewage works, and, to an extent, domestic consumers. Spring came at last, manufacturing resumed, exports began to go abroad again, but the episode did show the difficulties of a government's combining electoral support, humane compassion, and economic priorities. The press made much of the Labour Government's troubles.

Conclusion

Britain's war effort was greater, proportionately, than that of any other nation; postwar, unlike 1919–21, controls continued both to promote exports and to create a fairer society. Government was much more interfused in the economy, manufacturing industry's importance was acknowledged. Strikes which interrupted it became headline news. Britain in 1951 still looked rather tired and shabby, in need of a coat of paint, full of visible signs of patching up. The Festival of Britain in 1951 was an endeavour to provide something a little more colourful, lighthearted and encouraging. The war had been won, the slaughter ended, but it seemed a rather grey world. The prevailing mood was of dogged endurance, not of energetic enterprise, in spite of high employment and the considerable achievements of the economy in war and in peace.

Further Reading

Allen, G.C. (1970) *The Structure of Industry in Britain: A Study in Economic Change*, 3rd edn, Longman.

Floud, R. and McCloskey, D. (eds) (1983) *The Economic History of Britain Since 1700*, Vol. II, Cambridge University Press.

Hancock, W.K. and Gowing, M. (1949) *The British War Economy*, HMSO.

Pollard, S. (1962, or later editions) *The Development of the British Economy, 1914–1962*, Edward Arnold.

Questions

(1) What measures were taken to control the British economy, 1939–51, and how effective were they?

(2) Why was unemployment so low, 1940–51?

(3) Why did the size of firms and the degree of concentration change, 1939–51, compared with 1914–39 or 1870–1914?

(4) Why was there a strong demand for British manufactured goods, 1945–51?

8
Old Industries, New Industries, and Managerial Choices

Government intervention restricted managerial choice in old and new industries alike in these years. It had been intended that the Ministry of Supply would become the Ministry responsible for ordering every sort of supply and equipment for all the armed forces: instead, the Admiralty did most ordering for the Navy, the Ministry of Aircraft Production for the Air Force, and the Ministry of Supply was left to equip the Army and provide whatever non-specialist goods either of the other ministries asked it for. Similarly, ministries with control over civilian life proliferated, and a separate Ministry of Shipping both explained what shipping was available for imports and exports and allocated that shipping space when told what the essential imports and exports were.

However, to manufacturers the ministries sometimes seemed omnipotent. Most major industries had a Control, that is, a small committee or board consisting largely of men from within that industry who passed information, suggestions and requests to the appropriate ministry and who had to carry out the policy laid down by it, giving instructions to firms in their turn. They also operated inter-industry controls. It is understandable that, patriotic and hardworking as many managers were, they at times were exasperated at the form-filling and delays encountered. Equally, civil servants were exasperated at their explanations of delayed production.

Postwar, the Ministries of Aircraft Production and of Supply merged, and the Board of Trade took over control of most raw materials. Manufacturers were aware of three bottlenecks in production: shortage of labour, particularly of skilled men; possible interruptions to fuel and power supplies due to lack of coal and insufficient equipment; and delays in building new premises. The last was important, discouraging rethinking of factory and workshop reorganisation and layout. All building materials, most of them also wanted for the post-war housing programme, were subject to permits and allocations at least until 1948. This led to delays on building sites when timber or piping failed to arrive; and slow work was encouraged by the unforeseen effects of a bonus scheme for building workers. It was 1960 before the level of productivity in construction of 1939 was regained.

New tax laws encouraged manufacturing industry to invest — or so it was hoped. In 1945, 20 per cent of the cost of new plant was allowed to be offset against tax in one year — but the total allowed for depreciation was not increased. In 1949, the 20 per cent was doubled to 40 per cent, though total depreciation was still not increased. With company tax up to 50 per cent in some years, this offsetting was useful, but not necessarily decisive. Government thought industry had to be encouraged as well as controlled.

In 1945, the government set up tripartite working parties — government, management and union — to investigate and make recommendations for industries denied investment during the war. One result was the Industrial Organisation and Development Act (1947) which enabled tripartite industrial development councils to be established for individual industries, to fact-find and guide future development. Apart from the Cotton Board, these were short-lived and ineffective.

OLD INDUSTRIES

Textiles

Every textile industry, including rayon, had fewer employees and lower output in 1946 than in 1939, though all showed some improvement on 1946 by 1948. In 1951, textiles employed 1.1 million people, the postwar maximum. 1951 was also the peak postwar year for textiles as a proportion of exports, but whereas they had been nearly a quarter of all exports in 1937, by 1955 they were about a tenth.

Cotton

The cotton industry, built on the export market, provided 13 per cent of total exports in 1937, but only 4 per cent in 1951. Was this decline inevitable? Wartime limitations on the input of raw cotton, both because of shipping hazards and currency restrictions, of course decreased cotton's output. Its production was 'concentrated' in fewer factories and a third of its capacity was closed down; some was dismantled, so that war-related industries could use its factory space. Workpeople left for other, more lucrative employment, such as munitions manufacture. An Essential Works Order then prevented remaining workers from leaving without official permission. Once Lend-Lease was available, it was not so important that cotton should export. At home, much of its production was for the armed forces; its civilian goods were made to utility standards and their prices were controlled.

To maximise use of semi-skilled labour, the government tried to minimise closure of the small percentage of firms which used ring-spindles, but labour productivity did not noticeably improve. By 1944, fearing a further decline, the government sent the Cotton Textile Mission (Platt Mission) to the United States to study that country's cotton industry and its productivity. The resultant report was critical of British equipment, and of the industry's lack of vertical or horizontal organisation. It advocated vertical organisation and a reduction in the number of merchant houses involved in exports, since it believed that the latter, in trying to offer purchasers something a little different, decreased the chances of economic long runs of standardised cloth.

Relative costs changed, but little else did. Postwar wages in cotton were not high compared with those in other industries, yet employers considered them high compared with pre-war. The government, aided by further reports, accepted that, while automatic looms for weaving were still not cost-effective in British conditions (and in any case would have to be imported, using scarce currency, since few were made in Britain), ring spinning surely must be. The Cotton Spinning (Re-equipment) Act of 1948 tried to encourage both horizontal integration and ring-spinning by meeting a quarter of the cost of re-equipping with rings in large or closely associated firms. It did not have much effect. Manufacturers instead looked to immigrant girls and women, displaced persons in postwar Europe, who after a little training would work in existing ways for existing wages, paid according to the complicated lists which perpetuated an increasingly inefficient use of labour. Therefore the opportunity to re-equip at a time of good demand when, owing to labour shortage, it would have caused a minimum of social upheaval, was lost.

For a time, demand continued steady; by 1950 the British finishing trade was buying imported unfinished cloth, allegedly from Hong Kong, but much actually from Japan, because Lancashire could not expand production. The most damning indictment of Lancashire's inertia was seen in the Musgrave Mill experiment at Bolton. Without new machinery, but with work reorganised by the Cotton Board with the approval of both employers' and employees' associations and with an altered wage structure, output per man/woman-hour increased 39 per cent, weekly wages 30 per cent, and labour cost per pound of yarn fell by 10 per cent. Some changes followed this example, but not enough. Style, quality, and long-established preferences ensured some types of market at home and abroad, but the industry was not improving its position in an increasingly competitive world. Its labour skills in many types of yarn did not outweigh rivals' cheaper labour and increasingly sophisticated machinery. *Some* decline was inevitable: so much was not.

Woollen and Worsted

About 10–12 per cent of raw wool was British, and the rest came largely from non-dollar areas, especially Australia, South Africa and New Zealand, while Argentina supplied cross-bred wool. The constraint on imports of wool when war began was one of shipping space, rather than of currency. Exports of woollen goods were wanted to earn dollars. The industry had a greater number of middle-sized firms than did cotton; this made it easier for firms to avoid, and more difficult for the government to organise, 'concentration'. The industry's output was rationed and conformed to utility standards. The combination of controlled input, some 'concentration', compulsory redirection of labour, and voluntary movement of employees reduced numbers from 227,000 (insured employees) pre-war to 126,000 in 1944. Under war-time controls, all purchases of new wool were made by the government, then apportioned to the firms. Gains in productivity emerged from longer runs of standardised yarn and cloth under the 'utility' schemes, but these gains were offset by a rather high labour turnover. Government orders, basic civilian needs, and some exports were produced on ageing machinery.

Postwar, many controls continued. Purchase of wool had been in advance of use: two years' consumption was in hand. Demand was so strong that the store was dispersed in five years instead of the expected ten or twelve, yet prices of wool rose. Where in 1946 they were only 20 per cent above pre-war levels, by 1948 they were 300 per cent above them, and by early 1950, 400 per cent. At the time of the Korean War they were, at peak, 800 per cent of pre-war prices. Just as suddenly, prices then halved over the next six months.

Some British manufacturers had bought in advance over these years, foreseeing increases, and had passed on some of the increase to consumers. Others had hung back, and so were forced to buy at peak prices to continue production and then tried to pass on the higher costs. After the slump in sales (consumers had been pushed too far) most British firms weathered the storm, some by increasing use of non-wool fibres (though this was a small proportion), others by extended bank loans or by using the profits made abroad while the going was good.

Table 8.1 gives an indication of the industry's fortunes. Some re-equipment was done with the profits of the postwar boom, none of it very innovatory, except for fine knitting machinery in the worsted trade as the fashion for 'wool jersey' for women's dresses became popular.

Table 8.1 *The Woollen Industry, 1937–54*

Year	*Tops (m. lbs)*	*Yarns (m. lbs)*	*Cloth (m. sq. yds)*
1937	278.5	565.8	445.5
1949	286.4	528.6	438.6
1950	316.3	556.0	450.3
1951	241.4	501.3	418.1
1952	331.1	455.1	377.9
1953	322.9	534.9	411.8
1954	303.8	538.6	414.4

Source: Allen, G.C. (1970) *British Industries and their Organisation*, 5th edn, p. 283, Longman.

The Iron and Steel Industries

Throughout the period 1939–51, in spite of war, controls, nationalisation and denationalisation, about 40 firms produced iron and steel, with varying degrees of integration backwards into ore fields and forwards into steel processing. Dozens of smaller firms worked in assorted stages of production. When war began, the BISF's functions were largely taken over by the Iron and Steel Control, staffed predominantly by BISF men. It had powers of compulsion, but was itself subservient to the Raw Materials Committee of the Ministry of Supply, which consisted mainly of IDAC men who knew the industry well.

Postwar, the Control was replaced by an Iron and Steel Board which accepted most of the BISF's ideas for the development of the industry. Labour nationalised the industry in 1949, but not until September 1950 did the new Iron and Steel Corporation begin work. It did little but make plans before its hand was stayed by the incoming Conservative

Government of late 1951. All the changes at the top of the industry made little difference to its structure or development as an industry with many middle-sized firms, by international standards.

Wartime use of plant had to consider availability of inputs of all kinds — imported and domestic iron ore, scrap metal, semi-finished steel, coal, coke — and the shipping and railway rolling stock available for inputs and output. It also had to consider the vulnerability of East Coast works to bombing and this limited their development. Exports of steel were not used to gain dollars, but for quite specific purposes — to put around meat tinned abroad for British armed forces, or to build oil rigs in the Middle East to keep up the flow of oil to Britain. Imports of steel increased with Lend-Lease. It was sensible to import semi-finished steel (2.60 million tons at peak in 1942) since it was lower in bulk than ore and did not use coal in blast or steel furnaces and could be made into the exact shape required.

However, there were other inputs which had to be paid for, and since some were dearer than British products, levies on British steel ingots and on scrap were paid into a central fund from which payments were made to firms which had to use the dearer inputs. When the fund was exhausted, the government kept the payments going. In 1949, the industries again made payments to level out costs.

There was no great expansion of steel-making capacity during the war to burden the industry after it. Only four new blast furnaces were built, and one new steelmill, bringing capacity to 13 million tons. The government paid for some specialised plant which clearly would have no peacetime use, and it was dismantled. Iron and steel prices were controlled, and this was a spur to efficiency, since the more a firm could keep down its costs, the more profit it made. British domestic steel prices were higher than US steel prices during the war, but not, on average, higher than US steel plus its transport to Britain.

Postwar, since steel was such a key commodity in economic recovery and output, wartime controls remained, and allocation of steel for some time was a key ingredient in the government's endeavours to shape the economy. Capacity was expanded: 16 million tons were produced in 1950. While US tinplate and steel sheets were cheaper than British, other British products were lower in price; and all British prices were lower than most European prices. This was good news for British shipbuilders, car manufacturers and other firms using steel. When costs of iron and steel inputs rose with the demand caused by the Korean War, all countries were affected similarly and demand, which had shown signs of flagging in 1949, strengthened. British iron and steel was largely sold to the domestic market.

Most of the expansion of steel output was in additions to existing plant. There were technical improvements in some processes, assisted by what was learned of German practice, by the installation of some US equipment in Britain and by 'productivity' visits to the USA. But the sheer size of US plant, and the fact that there would have had to be an immediate increase in steel use in order to build plant on large greenfield sites told against adoption of US methods. Further, it was easier to process work, to overcome delays in delivery of building materials, to supervise construction, and to finance an addition to a works than to scrap and build. This decision also preserved the existence of firms reluctant to disappear in a wholesale reorganisation. One new venture, however, was a huge new rolling mill begun at Port Talbot in South Wales, financed and operated by the merger of Richard Thomas and Baldwins.

Many additions were sited to receive German scrap, a wartime legacy, bound to dwindle. A longer-term view might have dictated new plant on ore fields rather than on the coast or the Clyde. By the end of 1951, imported and home-produced ore replaced scrap. Steel was again subject to 'allocation' beause of defence needs, but it was allocation of a larger whole — a record steel output of 16.42 million tons in 1952. The dollar shortage continued to discourage imports, markets seemed assured, and the decisions to 'patch' rather than build anew seemed to have paid off.

The Shipbuilding Industry

Shipyard owners believed they had an acute shortage of labour early in the war: when orders picked up in 1938, they had taken on boys and young men as apprentices and trainees, and it was these younger people who left voluntarily or compulsorily in 1939/40 to join the armed forces. The Ministry of Labour thought owners had always been wasteful in their use of labour and that it was up to them to woo back some of the thousands of skilled older men who had left after years of unemployment. From 1939 to 1945 there was an uneasy balance between the availability of skills and materials for shipbuilding. Inter-union disputes still interrupted production.

There was no doubt about the need for ships. The Admiralty took over control of all shipbuilding from the Ministry of Supply in 1940 and gave priority to naval rather than merchant ship construction. In 1941, shipping losses were more than double the intended shipbuilding programme; from January 1942 to March 1943, only a quarter of the losses were made good by British yards. Output averaged 1.1 million tons a year, 1941–4, and fell slightly as men were taken away to work on the floating harbours necessary for D-day and the invasion of Europe.

Some of the gap between losses and output was filled by merchant ships built in the USA — 'Liberty ships' — standardised merchant vessels, prefabricated, welded together, and launched within a matter of days. US merchant shipping output rose from 1 million tons in 1941 to 19 million tons in 1943. Ships, complete with coathangers and electric clocks, were passed over to the British flag under Lend-Lease in 1943. British yards learned some lessons from American methods, although they did not have land available on which to construct large sections. However, smaller British yards prefabricated smaller ships from sections made in engineering firms up and down the country. Welding, now improved, was adopted. It produced a 10 per cent lighter ship, since metal plates did not have to overlap, nor need heavy rivets be inserted.

Postwar, the British merchant fleet was 4 million tons smaller than in 1939: 18 million tons had been sunk. Demand for replacement was high; so was foreign demand, and likewise demand for repairing and refitting liners which had been used as troop carriers. Employment in shipbuilding and marine engineering, 2.6 per cent of the total British workforce in 1935, rose to 4.4 per cent in 1948. By 1949, 220,000 were employed in the yards, and a further 80,000 in marine engineering. Output rose to over 1.5 million tons per annum and by the end of 1949, a postwar peak 45 per cent of world tonnage was under construction in British yards. After some decline in orders, the Korean War brought a resurgence. British yards were stretched to capacity. Japanese and German yards, slowly rebuilding, were producing for home consumption; other countries were subsidising and increasing output for themselves.

The flow of orders continued, due to the comparatively low price of British steel and the skill of the British workforce. Although there were the beginnings of further changes — some shipyards built in covered areas where, in all weathers, pre-assembly could be done, and bigger cranes were installed for lifting heavy sections into place — their organisation remained much the same. Higher profits were made with higher prices, but the increases were comparatively modest, and not devoted to re-equipment. Shipbuilders preferred a long order book, encouraged by comparatively low increases in price, to rapid profits followed by a possible dearth of customers. For the time being, the industry looked very prosperous; this bred a complacency about future competition, and a sense of urgency about the completion of the orders was lacking.

The Engineering Industry

Much of the old output of engineering virtually ceased, 1939–45, but there was an enormous increase in newer engineering products of many kinds. Machine tool firms expanded output at first by employing subcontracting firms, and then by enlarging. At the height of demand in 1942, output of machine tools was worth £33.54 million, produced by 95,780 people. Postwar, much hope for increased productivity turned on the use of machine tools, and the machine tool industry itself became rather more productivity-minded. In 1951, 53,210 workers produced £47 million of machine tools, a considerable improvement in productivity (even allowing for price rises). Exports varied in volume: from 1946 to 1951 between a half and a fifth of output was sent abroad, largely to the Commonwealth. At home, machine tools were more widely installed: a total of 28,749 in use in 1947, and 41,171 in 1955. Their rate of installation was affected by changes in tax policy. A backlog of orders built up; the employers blamed the delays on skilled men leaving the industry, but seemed unwilling or unable to state convincingly to the government or trade unions that exceptional wage rises might be necessary to retain men in this seedcorn industry, or to speed production by standardisation. Instead, much attention was given to the industry's selling organisations, largely agencies which kept the machine tool makers at one remove from their customers, unlike their more forceful American counterparts.

Two of the most important sections of the armaments industries, aircraft and shipbuilding, are dealt with elsewhere. Here, attention is given to tanks and guns. Manufacture of guns of all kinds was divided between the publicly-owned Royal Ordnance Factories (ROFs), private armourers such as Vickers-Armstrong, and firms with no armament experience. The proportion of work done by private industry (both traditional armourers and 'outside' firms) and the publicly-owned ROFs varied with the type of equipment.

Army weapons had been neglected interwar, and this became especially evident in tanks, where early designs were underpowered and under-armed. Not until 1945 were suitable tanks made. Car makers had tried to manufacture them, but tanks and cars were very different: there was more similarity to heavy trucks, but even truck builders such as Leyland and Vauxhall had difficulties building tanks. Locomotive builders also tried and failed. At first all tank production was in private hands, with small scattered teams working on design, but later ROFs were also used. There was an insufficient pool of design and production engineers on which to draw.

Production of textile machinery practically stopped until 1945.

Postwar, however, it was in great demand for the re-equipment of the cotton industry. Ring spindles were produced for the home market, but for weaving, old 'Lancashire'-type looms were made in quantity to make up deficiencies quickly. By 1951, total employment in the manufacture of textile machinery stood at 75,000. Locomotive and rolling stock production also virtually ceased, 1939–45, then gradually resumed, by which time overworked engines and wagons were in urgent need of replacement.

NEW INDUSTRIES

The Engineering Industry

Most of the light engineering industry was turned over to war production, 1939–45, making every kind of essential small item. Postwar, the growth in car production stimulated the engineering component industry. Office machinery and equipment for the home found domestic and export markets. The clock and watch industry which had almost disappeared from Britain was re-established with a government subsidy, because of the strategic importance of clock-like mechanisms in fuses and measurement devices in aircraft.

Motor Vehicles

The few cars produced during the war were for military purposes and so were most lorries. Motor vehicle firms made tanks, aero engines and aircraft. Reconversion to peacetime production was therefore a major operation. In 1946 car production was no more than half its 1937 figure; by 1949, the prewar peak of production had been surpassed; Table 8.2

Table 8.2 *Cars produced in Britain, 1936–1950 (000s)*

Year	*1936*	*1937*	*1938*	*1939*	*1945*	*1946*	*1947*	*1948*	*1949*	*1950*
Car production	354	390	341	305	17	217	287	335	412	523
New registration (excl. taxis)	310	318	280	236	8	122	148	113	155	133
Export (complete cars)	51	54	44	—	—	70	126	195	219	344

Source: Maxcy, G. and Silberston, A. (1959) *The Motor Industry*, Appendix D, Tables 1, 2, 4, Allen and Unwin.

shows the figures. In 1950, Britain exported over 70 per cent of production, half of all world exports of cars. There were four reasons for this: pent-up demand for cars in world markets; restriction of car sales in the home market through a high purchase tax; allocations of raw materials, including steel, made to car firms provided they manufactured for export; and lastly, postwar US cars were too big, too petrol-thirsty, and cost too many dollars for European — and other — markets. German competition had been much diminished by the effects of the war.

It was estimated that in 1947 over 400,000 were employed in the motor vehicle and components industries. Thirty-two firms made cars in 1945, but the so-called 'Big Six' were dominant — four British (Morris, Austin, Rootes and Standard-Triumph) and two American (Ford and Vauxhall). Two thirds of the factory cost of a car were 'bought in' from component manufacturers in 1946. Dunlop Tyres, Lucas, Triplex (glass) and GKN (engine and transmission components) between them supplied two fifths of all components. Because all effort for six years had gone into war-related work, the cars of 1946–51 were technically much the same as prewar; the production lines and flow of production were identical to those of 1939.

A higher proportion of commercial vehicles than of cars was retained for the home market — some 45 per cent, 1948–50. When railways and road haulage were nationalised under the 1947 Transport Act, there was still some administrative discrimination against road and in favour of rail. Nevertheless, almost *all* the postwar increase in internal transport in Britain was an increase in road transport, particularly in light vans.

Petrol was rationed for private vehicles until 1951, to limit expenditure on oil imports; Britain was thought to have no indigenous oil supply. Since motor cycles used considerably less petrol than cars, a higher proportion — 53 per cent — was allowed to be sold on the domestic market. Table 8.3 gives output figures for motor vehicles other than cars and taxis. For employment and for exports the motor vehicle industry was enormously important. It was a seller's market but, in spite of some grumbles, the buyers appeared to be content with what they got.

Aircraft

At wartime peak, over 1.5 million people worked on Ministry of Aircraft Production contracts, making everything from aero engines to parachutes in large modern factories (often shadow factories) and in small old-fashioned subcontracting firms. The problem was to coordinate production in a hugely expanded industry through administrative

Table 8.3 *Motor Vehicle Output, 1939–1951 (000s)*

Year	*Commercial vehicles*	*Agricultural & industrial tractors*	*Motor cycles*
1939	97	10	n.a.
1945	122	18	n.a.
1946	148	28	93
1947	157	56	112
1948	177	114	131
1949	218	88	153
1950	263	117	171
1951	259	137	172

Source: Allen (1970), p. 177.

means rather than the market. US machine tools and components were essential to output.

In 1940 aircraft were built at maximum speed, without a thought-out plan; sustained and planned aircraft production (especially of bombers) in much larger numbers followed: 26,461 were built in 1944. New radar and electronic navigation equipment was added to aircraft; engines became increasingly powerful through continuous modification and through a major radical innovation, the gas turbine or jet engine, first used operationally in 1944.

Jet engines, most widely produced in Britain (the only ally to make them) were not uniquely British; Germany also had them. In Britain, Frank Whittle developed them from a mid-thirties prototype made in his own firm, with minimal Air Force interest. Only Rolls-Royce initially could work with enough precision to make jets well; by the end of the war four firms made jet engines in projects in which government and industry collaborated. Initially built with centrifugal compressors, jets looked quite different from sleek modern engines which work with axial compressors.

Subcontractors accounted for about 40 per cent of the output of the industry: most had no previous connection with it. Large boiler makers and small chemical works were pressed into service. Large firms such as English Electric, with no past experience, made aircraft to designs drawn up by the principal aircraft firms. They used unskilled and semi-skilled labour in large numbers; the aircraft industry was more labour-intensive than the German or US industries. A shortage of good managers meant that the Ministry of Aircraft Production often directed work to well-managed firms rather than to those in other respects superior. In 1943 Short Brothers was nationalised and its management reformed; a government controller took over Fairey Aviation. These measures indicate serious faults in past management. As aircraft became more complex, individual designers were replaced by teams of highly qualified specialists in aeronautics in the USA; in Britain these specialists, excellent but in short supply, were spread too thinly amongst the aircraft firms.

Peace meant a quick, massive contraction of some 80 per cent in the aircraft industry. The government then gave contracts to all major wartime producers to keep a number of design teams in being. Output of standardised aircraft of the latest design could be increased quickly in emergency, but the policy did keep talent dispersed. Many postwar jet fighters found large overseas markets: Britain had a lead in this kind of technology due to Whittle.

Nine types of civil plane were suggested by the wartime Brabazon Committee; one of them, the Comet, the world's first jet airline, was flawed by hitherto unknown metal fatigue; another, the Viscount, was Britain's most successful airliner, postwar. A third, the Bristol Brabazon, named after the Committee's pioneer aviator chairman, was heavy and unwieldy: only two were built. The government specified and paid for the prototypes. Interim types of plane were also ordered. They were not new designs; some were converted bombers. Noisy and uncomfortable, they did not sell well or enhance the reputation of British civil aircraft. The best export sales were of the de Havilland Dove (409) and the Vickers Viscount (356), both Brabazon types.

The problem was to reconcile maximum technical achievement, total reliability, and economy of production and operation in civil aircraft. The aircraft industry's output was estimated at £136 million in 1949, of which £85 million was for military aircraft for government, £23 million for R & D (research and development) for government, £20 million for export sales, and £8 million for sales of civil aircraft on the British market. It needed big sales of any one model to justify the initial outlay on R & D and capital equipment, and the British home market was small.

Electrical Engineering

Some increase in demand for electricity was met by more intensive use of existing generating equipment. Additions to generating sets were piecemeal and minimal. Postwar, a surge in demand occurred. Difficulties were shortages of steel and concrete needed to construct new power stations or add to old ones to house new sets, and delays in turning out turbines, boilers and other equipment in sufficient quantity, yet with specifications which differed according to the taste of engineers of various power supply companies. Nationalisation of the electricity supply industry in 1947 made easier the decision of the Ministry of Supply only to permit sets of 30 MW or 60 MW turbo-alternators. This speeded production, but steel shortage was still a limitation. Interruptions to electricity supply, caused by overloading the system and by shortages of coal, did not deter would-be users of electricity. The National Grid proved a boon; extra transmission lines were laid during the war, enabling power to be sent from blacked-out London, where electricity consumption fell, to factories in the North and Southwest.

Much in demand for everything from high-powered transmission lines to the increasingly complex instrumentation of aircraft, cables were in short supply during the war. The Cable Makers Association, which kept a tight grip on prices and imports in peacetime, had to admit defeat, and emergency cadmium-copper lines, instead of the usual aluminium and steel, were imported from the USA, together with steel pylons in sections. Rubber substitutes replaced natural rubber as an insulating material as the war progressed. Postwar, two major cable firms merged to form British Insulated Callendar Cables Ltd (BICC), thus strengthening research possibilities, but also further limiting market competition.

Light bulb production expanded from over 172 million in 1945 to over 252 million in 1950. The subsidiaries of AEI and GEC dominated output, and retained their international agreements on markets and prices.

Coal, the usual means of heating British homes, was rationed until 1958; cheap (£1.50) electric fires plugged in to the mains were an obvious alternative, even if running costs were comparatively high, and their use (at peak in the early morning and evening) a great burden to the over-stretched electricity supply industry. With well over one million made and sold per annum, a steep purchase tax on their sale and a severe limitation of the steel required for their construction then limited production to 70,000 per annum for some years.

Electric water-heaters and electric cookers, aided by hire-purchase and rental schemes, also became more popular; electric cookers increased from one tenth to one half of all electrical appliance sales between 1947 and 1957. By 1946, 38 per cent of electricity consumers had a vacuum cleaner but refrigerators and washing machines, owned by 2.2 and 2 per cent of electricity users, were still unusual. Almost all these appliances were made in Britain; some were also exported.

Electronics

By 1939 Britain was protected by a chain of radar stations along its coast. In 1940 scientists working on radar at Birmingham University developed the cavity magnetron, a new type of wave generator put into production by GEC. Cathode-ray tube production was increased to meet the demand for radar displays and test equipment though some US imports were needed too. It was of course essential to the production of television sets, resumed after the war. Military electronics remained of great importance. Its highly successful postwar development might have been at the expense of civil electronics, yet because requirements of radar were much more exacting than those of previous civilian work, technical capabilities were substantially and permanently raised. Some firms not hitherto involved in the electronics industry proper entered it and stayed on, postwar. The Decca Record Company used its wartime experience to develop the Decca Navigation System for ships, a postwar best-seller.

A number of firms, usually linked to the large electrical companies, made valves and tubes and also produced radio and television sets, or were associated with companies that did. Only 0.2 per cent of electricity users owned a TV set in 1946, but 34.7 per cent did so by 1955; most sets were British-made.

The Chemical Industry

A whole range of products, from high explosives and propellants to poison gas (never used in World War II) and light alloys for aircraft, was made in chemical plants. The chemical industry in general, and ICI in particular, was well prepared for war, both technically and in productive capacity — very different from 1914. Thirty-five of the government's chemical 'agency' factories were managed by ICI, the government's largest agent.

New products built on the scientific and technical developments of the 1930s. In the 1940s the basic materials for organic chemistry were coal and molasses. The latter was fermented to produce organic chemicals, principally by the Distillers Company, and was the base for about 30 per cent of organic chemical production in 1949. The most important developments were new synthetic plastics and fibres, some oil-based, and new pharmaceuticals and pesticides. Polythene was discovered accidentally by ICI research scientists in 1933, looking for a new plastic material. Polythene was produced because the gases used in a high-pressure experiment were impure! For many years ICI did not see polythene's potential. When production began in September 1939, its main use was as an electrical insulator for cables, particularly for radar sets, permitting airborne sets to be much lighter. Another ICI plastic was 'perspex', used in aircraft windscreens. Production increased from 357 tons in 1939 to 4,700 tons in 1944. The third important plastic was PVC, a German (I.G. Farben) discovery developed from vinyl chloride, brought into production by ICI in 1940 as a rubber substitute in electrical insulation.

British R & D in plastics was small and late by international standards, as may be seen in Table 8.4.

Table 8.4 *Beginning of Commercial Production of Certain Plastics*

	UK	*Germany*	*France*	*Italy*	*Japan*	*USA*
PVC	1940	1931	1940	1951	1939	1933
Low-density polythene	1939	1944	1954	1952	1954	1941
Polystyrene	1950	1930	1951	1942	1957	1933
Polymethylmethacrylate (Perspex)	1933	1930	1938	1937	1938	1936
PTFE	1945	1958	1959	1955	1963	1943

Source: Freeman, C., Young, A. and Fuller, J.K. (1963) 'The plastics industry: a comparative study of research and innovation', *National Institute Economic Review*, No. 26, p. 41.

Heavy chemicals — chlorine, caustic soda, soda ash, and various industrial acids — still formed the bulk of production of the chemical industry. In sulphuric acid production the contact process became firmly established. In nitric acid production (wanted for explosives), too, there was a consolidation of production using new processes. The increase of nitric acid output from 1949 was due to government subsidies to fertilisers for agriculture.

Biologically active chemicals, such as pharmaceuticals, pesticides and weedkillers, react on living matter. In the 1930s it was discovered that certain organic compounds could be used to treat bacterial infections. British companies, including May and Baker and ICI's Dyestuffs Division, began work on this. In 1942, ICI set up a separate pharmaceutical company and discovered an important new sulphonamide drug, sulphamezathine. The most famous anti-bacterial compound, however, was penicillin, put into large-scale production in Britain and the USA. Work on it (and other drugs) was shared in Britain between Glaxo, Beecham, and the Distillers Company. Quinine was used to combat malaria, but quinine plantations were lost during the war and synthetic drugs had to be developed. The first were German, but a research programme led ICI to produce a superior drug, paludrine. Pest control was advanced by the Swiss discovery of DDT, and ICI's development of gammexane as an insecticide. Hormonal weedkillers were also developed, with ICI playing a leading role here, too.

Plans were made to expand R & D considerably postwar, and a Central Laboratory to engage in fundamental research was established. Between 1938 and 1952, ICI's research expenditure increased by over 600 per cent.

Synthetic Fibres and Textiles

Imported wood pulp for viscose rayon was given some priority in shipping in the early years of the war, and Courtauld produced a little under 90 per cent of its 1939 output in 1945. In March 1942, one ministry was about to remove wood pulp from the priority shipping list just as another was urging the industry to make more rayon fabric under the utility scheme. The latter prevailed. Viscose yarn, used instead of cotton in car and other tyres in the USA, was used in British tyres too, and at peak, in 1943, Courtaulds sold 260,000 lbs of it a week to the government for this purpose.

Postwar, rayon was cheaper than cotton. Exports of rayon filament were double their 1937 volume in 1950. Staple fibre had even more growth: 6 million lbs of exports in 1937, 22 million lbs in 1950, but 15 million in 1951. Viscose rayon was not as suited as rayon acetate or the new crease-resistant cotton to the longer, fuller skirts of the 'New Look' of the 1950s, but still had a market.

New yarns were made from acetic anhydide and acetone, provided by Distillers, which derived it from molasses. A similar yarn could be part of the product of an oil-cracking plant, and British Celanese was the first British firm to build one and use the product. Nylon, made in the USA for some years, was produced from either coal or oil. In 1940, ICI obtained patent rights from DuPont (USA) and, jointly with Courtaulds, established British Nylon Spinners. All its output during the war was used for cloth for parachutes. Postwar, nylon was more commonly knitted than woven, and nylon stockings and underwear, well publicised for their glamour, were eagerly bought. In 1939, the Calico Printers Association laboratory discovered Terylene. Its development was postponed by the war; ICI acquired its patents, and began large-scale production in the late 1940s.

The newer fibres, knitted or woven, were more resilient and quicker-drying than cotton, and creased less readily than rayon. Utility fabrics, including rayon, were free from purchase tax until 1952 and this helped the industry. Productivity gains were made, but not spectacularly, and to help export sales, the Rayon Producers Committee, which became the Man-Made Fibres Producers Committee, put a levy on home sales to assist exporters.

Courtaulds in 1945 found that nearly 50 per cent of its looms were made in or before 1920, and it replaced many and also invested in a new continuous process for filaments which meant more (British-made) plant. The firm established a proper research division, which provided both its future chairman and its future products. Manifestly, it was a firm affected by price change, by fashion, and by competing fibres. Encouragingly, it seemed prepared to acknowledge and plan with this in mind. Its US subsidiary, AVC, was sold off during the war by the British Government, using its emergency powers, to give backing to the claim that Britain was desperately in need of financial help from the USA. AVC's profits and productivity were not good by American standards, and the price fetched was low. Its sale was, and was intended to be, symbolic of Britain's dire financial state.

Ready-Made Women's Wear

Factory production of men's clothing was well-established between the wars; women's clothing was made predominantly in small-scale workshops, or sewn in the home, or by a dressmaker of neighbourhood, local, or even national renown. A few factories made good-quality tailored clothes for small, fairly expensive shops or for large department stores, but much of the cheaper output of small firms was rather shoddy. About 100,000, mostly girls and women, were employed in the industry.

Wartime regulations changed the industry to a factory-based one and also altered its location. 'Concentration' schemes diminished establishments in the Midlands, Leeds, and Manchester, and expanded them in London and Scotland. Larger firms, and those small firms which did contract work for them, found their work simplified and speeded up by government regulations which, from 1942 onwards, restricted the use of trimmings, limited the amount of material used in width or length of skirt, and confined each establishment to a small range of styles. This trend to fewer factories and more standardised output meant that firms took better advantage of electric cutters and presses. The Ministry of Supply encouraged a better flow of work through the factory by technical advice and, in a seller's market, clothing firms were no longer affected by retailers' seasonal sales. Additionally, utility cloth (used in 'utility' clothing) was of at least a minimum standard so that cutting and sewing it was easier. The consequence of these changes was that productivity per worker rose between 25 and 75 per cent.

Clothes rationing continued until 1949, though with rather more items available outside the ration scheme; non-utility items were subject to a 66⅔ per cent purchase tax. Exports were wanted and government orders for the forces fell. Demand was steady, and pre-war fluctuations were in abeyance. A further change was in the greater sale of brand-name clothes: the maker, rather than the place of sale, became the assurance of quality to the purchaser. The 1951 increase in textile prices affected the industry badly, however, and shops once more held 'sales' to clear stocks, and meantime reduced their orders to manufacturers. Manufacturers, especially those who after 1945 expanded into premises better equipped, but therefore with more expensive overheads, often in Development Areas, felt the pinch. Nevertheless, the outcome was not a retreat from factory-organised production. Once the utility scheme ended in 1952, there was more attention to quality than prewar and more to fashion than in the period of government restriction. A good deal of the distinction between middle-class and working-class dressing disappeared, aided by the growth of women's magazines with a wider circulation than prewar. Ordering by multiple stores also became more important to the industry. Subsequently, new, smaller firms began to appear, especially in London, and in the 1960s, the city was the centre of the popular and inexpensive fashion world.

Manufacturing and Research

War accelerated technical development in industrial sectors most closely allied to war needs, and postwar this development had some spin-off. Much technical and research expenditure was quite beyond the financial safety limits of individual firms or even industries. Therefore the government funded research and development both in its own establishments and in private-sector firms. Much of it was in aerospace, electronics and nuclear energy, and this expenditure continued into peacetime. The prestige of science in the eyes of the government grew: the boffins (a term first applied to scientists working for the Air Force) had come to stay.

Yet in some industries, as has been seen, insufficient attention was paid postwar to applied science, to technology, to developmental and production engineering. This helps explain the slowdown in productivity growth, and the failure to match first-class achievements with sustained output of quality products.

Conclusion

There can be no doubt that leaving the economy to the free play of market forces during the war would have led to quite unnecessary expenditure (since rival government ministries were the end purchaser of all war goods), to delay in removing bottlenecks in production, to inadequate provision for military needs and the civilian minimum, and quite possibly to military defeat. By 1942, manpower was the commodity in shortest supply, and so it was not always possible to organise shift working to make the most of equipment, neither was there the money to re-equip with less labour-intensive machinery, so there were piecemeal rather than large-scale alterations to methods of production. The need to export in order to pay for war materials of all kinds and for foodstuffs was vastly reduced by Lend-Lease. Only about one sixth of all imports was actually paid for.

Postwar, in spite of US help, most imports had to be paid for. Foodstuffs, raw material for civilian products, and inputs into goods that ultimately would be exported, were all necessary imports. Had there been fewer or no government controls, domestic demand from the better-off might have swallowed up the resources needed to provide the exports which paid for the imports. It would have been the poorer people of Britain who would have suffered most. Apart from considerations of what was socially desirable, the cooperation necessary to sustain production postwar might then have crumbled (this is discussed in the next chapter). No-one would have benefited from this. Given the postwar condition of the British coal and steel industries, it is highly unlikely that uncontrolled competition for their products could have caused greater production. Inputs for industry were decontrolled just as soon as they were plentiful enough to forestall bottlenecks and/or price rises on a significant scale. The same was true of food rationing; bread was not rationed until after the war, when wheat imports were restricted, and other items were at times reduced, but rationing ended when supplies became more plentiful. This helped to restrain price rises.

It is true that by 1951 the rate of investment in Germany was higher than in Britain, but the differences cannot be explained by the presence of controls in Britain. When the controls ceased, the differences remained. Nor can they be explained by the British Welfare State. Expenditure on housing, education, child care, school meals, welfare services, health services, national insurance, pensions, and social assistance was 16.5 per cent of GNP in 1950, about double the pre-war proportion, but most nations had increased their social outlay. However, British military and defence expenditure, though lower in 1950 than in 1938 in constant prices per head, was of course higher than in the defeated and virtually disarmed countries and, because of the Korean War and the imperial legacy, almost doubled by 1952, and this undoubtedly absorbed some of Britain's physical and psychological resources.

Demand was so strong that for the time being Britain's labour-intensive mode of production did not seem a handicap. As the next chapter will show, wartime agreements to alter work practices were conditional on a return to them postwar. Perhaps the most valid criticism of government controls is that they helped perpetuate in management an attitude of battling through, rather than one of fresh venture and energetic reappraisal.

Further Reading

Allen, G.C (1970) *British Industries and Their Organisation*, 5th edn, Longman.

Calder, A. (1969) *The People's War, Britain 1939–45*, Jonathan Cape.

Hancock, W.K. and Gowing, M. (1949) *The British War Economy*, HMSO.

Pollard, S. (1962 or later editions) *The Development of the British Economy, 1914–1962*, Edward Arnold.

Questions

(1) Compare and contrast government intervention in two of the major industries discussed in this chapter, 1939–51.

(2) What accounts for the differences in the fortunes of the cotton and wool textile industries, 1939–51, or 1945–51?

(3) Why was ICI such a successful company?

9

Employees, Trade Unions, and British Manufacturing Industry

The possibility and then the reality of war brought government and union leadership together, at first in uneasy consultation, and then in more assured cooperation; agreements marked a new attitude towards, and a new role for, trade unions. In the workplace, employee/management relationships were conducted largely within the bounds set by government and trade union leadership.

In August 1939, the government set up a National Joint Advisory Committee (NJAC), consisting of 15 employers and 15 trade unions, to advise the Ministry of Labour on matters affecting the workforce. Until May 1940, it accomplished little. Action began when the new Prime Minister, Winston Churchill, appointed the TGWU's Ernest Bevin, the outstanding trade union leader (and subsequently a Labour MP) to be the new Minister of Labour and National Service. Different as the two men were, they liked each other, united by the conviction that the defeat of Nazism was the overriding objective. Bevin, in this, reflected the views of most trade unionists. He got the NJAC to establish a Joint Consultative Committee, a smaller, more workmanlike body, which he consulted and whose advice he took. The government could force almost anything under Emergency Powers Acts (1939 and later), but he wanted cooperation, not coercion. When the Committee proved hostile to wage regulation, he dropped the idea; when it agreed to a National Arbitration Tribunal, he had this incorporated in Order-in-Council 1305, the Conditions of Employment and National Arbitration Order. Bevin was the propelling power behind the revised dilution agreements of 1940, and to smooth the way he promised and got through Parliament the Restoration of Pre-War Practices Act (1942) which promised to do just that when hostilities ended.

Other national bodies, including the Central Production Advisory Committee, also had trade union, employer, and government membership. Under the wing of the Ministry of Production, and on the advice of a small committee chaired by Walter Citrine, secretary of the TUC, national and regional tri-partite (i.e. government, employer, and union) production boards were set up. Prodded by the shop stewards movement, joint production committees including employers and employees were established at individual factories and plants. The shop stewards, regaining strength as factories became fuller and busier, contained a considerable communist element, and had been hostile to participation in the war (though the real threat of invasion in 1940 subdued them) but, when Russia was invaded by Germany and entered the war as an ally, the shop stewards in a national conference called for greater efficiency and higher productivity in munitions factories. The joint production committees increased their importance.

Employers reacted to the committees with varying degrees of suspicion and enthusiasm. In private, the men in the Ministry of Aircraft Production groaned, afraid that the committees might simply encourage grievances and delay output, but when Stafford Cripps became Minister, he welcomed them. The committees, first introduced into ROFs, worked well on the whole. Coal mining, rather than manufacturing, was the focus of disruption.

The Labour Party won the 1945 election on a programme supported by the TUC and had union-sponsored MPs among its twenty Cabinet members. Its nationalisation programme did not mean worker control of those industries, though consultation was built into their organisations and trade union leaders were appointed to their boards (although unions required them to resign their union positions on accepting these appointments). The government repealed the 1927 Trade Disputes Act, with its 'contracting in' clause. A National Joint Advisory Council for Industry continued the work of Bevin's wartime Committee, and some regional boards remained active. Trade unionists were included in teams sent by the government to survey the wonders of American productivity. Thus the close organisational and personal links between government and union leadership remained. Both wartime and postwar governments continually pressed for more output and longer hours, and hoped for wage restraint, and in this too union cooperation was obtained. Part of the reason for this was in the unanimity of war aims; and postwar, as Lend-Lease abruptly ended, trade union leaders backed the government's call for exports and continued controls. Most cooperated again during the fuel crisis of 1946/7. With their approval the Restoration of Pre-War Practices Act was temporarily suspended.

However, it would be quite wrong to assume that there was no friction at grass roots, or disagreements at the top. There were strikes in coal mining, shipbuilding, and on the docks, in spite of official trade union opposition. When, postwar, Russia rejected Marshall Aid, and the 'cold war' divided Europe, communists in various unions resumed their agitations, though the TUC urged unions to remove them from office. The first significant break in trade union/government cooperation however came through disagreement over wage restraint, and the break was recognised by the Labour Government withdrawing Order 1305 in August 1951. The apparatus of compulsion had only been effective while a cooperative spirit lasted.

Trade Unions, Workers, and the Law

In 1938, government and trade union leaders agreed that, should there be war, no compulsory direction of manpower, not even into the armed forces, would take place, though people in reserved occupations might be required to stay in them, to avoid skill shortages. Nevertheless, conscription was introduced, and civilian workers and their employers were increasingly controlled. The Restriction of Engagement Order of June 1940 made it obligatory for employers to engage workpeople only through employment exchanges or recognised trade unions. Its aim was to prevent 'poaching' without resorting to the unpopular leaving certificates of World War I. The Essential Works Order of 1941 gave government power to direct men and women to particular workplaces. Factories deemed 'essential' had to observe standards of welfare and have reasonable wages, but workers could not leave them, or be sacked, without permission from the Ministry of Labour.

The two laws were used sparingly, particularly with women. Postwar, they were retained, but used only in coalmining and agriculture, both short of labour, and both nationally important. They were ended in 1950, because they had become a reason for people not wishing to begin work in the industries which they covered.

Order 1305 made arbitration by the National Arbitration Tribunal compulsory if all else failed. Strikes and lockouts were illegal, except where the Ministry of Labour failed to intervene; employers had to accept wages and conditions agreed by collective bargaining for particular industries and districts. There were more strikes than in 1914–19 (or 1927–39), but they were shorter. There were only 107 prosecutions under Order 1305 against strikers, and only two prosecutions of employers. Postwar, there were no prosecutions until 1950, though strikes increased. The Order's 1951 withdrawal followed an unsuccessful prosecution.

The Growth of Trade Union Membership

By the late 1940s over 40 per cent of potential trade union members were members of trade unions. Membership grew both through active recruitment amongst new workers and by rapid enrolment in industries where, for the first time, employers recognised unions' rights to negotiate for their members, as Table 9.1 shows.

Table 9.1 *Trade Union Membership and Density of Unionisation, 1938–1951*

	Membership (000s)	*Density (%)*		*Membership (000s)*	*Density (%)*
1938	5,842	30.5	1945	7,875	38.6
1939	6,053	—	1946	8,803	—
1940	6,613	—	1947	9,145	—
1941	7,165	—	1948	9,319	45.2
1942	7,867	—	1949	9,274	44.8
1943	8,174	—	1950	9,289	44.1
1944	8,087	—	1951	9,535	45.0

Source: Price, R. and Bain, G.S. (1983) 'Union growth in Britain: retrospect and prospect', *British Journal of Industrial Relations*, Vol. 21, pp. 46–48.

In manufacturing industry unionisation was greater than for employment as a whole and density of its unionisation in 1948 was about 51 per cent overall (58.2 per cent for manual workers, 13.2 per cent for white-collar workers). The TGWU, with over 1.25 million members in 1949, was the largest union; the AEU, with over 700,000, was the largest in manufacturing.

Inter-union disputes continued. The TUC's remedy was voluntary federation of unions, but unions were not inclined to this, and so old rivalries remained.

Working Hours

Shortage of labour and intense pressure to maximise output of munitions led to calls for the maximum possible use of overtime by 1940. At the height of the Battle of Britain, a seven-day working week was urged by Beaverbrook for aircraft workers. Such a pace could not be maintained. Workers were exhausted and the quality and quantity of work suffered as a result. The lessons of World War I had to be learnt all over again. More than 52 hours' work a week actually reduced output. Yet in spite of the stresses caused by long hours, strikes in 1940 lost fewer working days than for 50 years.

After World War II, as after 1914–18, hours of work in many industries were reduced by collective bargaining, not government regulation. In engineering, the standard working week fell to 44 hours; in cotton it dropped from 48 to 45. The forty-hour week once more became the goal, but not yet the reality. Behind the pressure for reduction in hours was a sense of just reward for wartime effort.

Wages

Remarkably, during the war and postwar years there was no compulsory incomes policy: wages were determined by collective bargaining, although much modified by government intervention through compulsory arbitration (and taxation). Pressure for wage increases was reduced by subsidised foodstuffs and price controls accompanied by rationing. Income tax was extended to levels of income where it affected industrial workers in substantial numbers. The government's careful manipulation of prices of goods in the cost-of-living index helped prevent large wage demands. Wages increased from 1939 to 1944 by about 11 per cent above the index but earnings, which of course included any overtime above a 47- or 48-hour week, were considerably higher.

In January 1947 the Labour Government published a White Paper, the Statement on the Economic Considerations Affecting Relations between Employers and Workers. It called for wage restraint and was backed by the National Joint Advisory Council. It had little effect and a year later Attlee, the Prime Minister, without consulting trade union leaders, called for an indefinite freeze on wages, salaries and dividends. It was a notable tribute to the authority of the Government and of the TUC leadership that, at a special conference of trade union executives, the wage freeze was accepted in principle despite opposition from union leaders representing low-paid workers. (To soothe them, the TUC contradictorily stressed the need also to raise lowest wages.) Even at the height of the war no such intervention had been attempted. The wage freeze marked agreement that exports were necessary, that wage increases would increase the cost of exported goods to a point where they could not be sold abroad profitably, tacit agreement that wage restraint was necessary until more goods on which to spend wages appeared, and that wage restraint was at least an important ingredient in maintaining full employment in the future.

After devaluation, the TUC accepted a wage standstill, provided that the cost of living rose no more than 5 per cent. In fact it rose only 2 per cent. Then, from the end of 1950 to the end of 1951, retail prices climbed 12 per cent, while hourly wages and hourly earnings rose 10 per cent. Consensus on wage restraint ended and some employers, anxious to pay more to secure labour which was still scarce, were not sorry. Trade unions were beginning to look abroad at percentage wage increases, and this added impetus to wage demands.

Over women's pay, the unions were in a dilemma. They wanted to distinguish between traditional 'men's work' and 'women's work', but could not allow women to receive less pay than the men they had replaced, since this would lead to the undercutting of men's pay in due course. They therefore insisted that after a period of training such women should receive the full rate for the job. Yet in general women received lower wages than men, partly because men were given more senior positions, and this led to some strikes. Demands for equal pay for women became so insistent that the Government was forced to appoint a Royal Commission in 1944. It did not report until the war was over, and by that time most women had left their war jobs and the question of equal pay was no longer a live issue. It was not until 1975 that equal pay for equal work became a legal requirement.

Conditions of Work

There was significant improvement in working conditions in many factories, 1939–51. Shadow factories incorporated the latest facilities. There were improvements in factories covered by the Essential Work Order, and factories where women were employed on war work had to provide washrooms and other health and safety facilities, and men benefited from these improvements too. Canteens were introduced. Many changes were permanent: postwar, the motor industry had some of the best working conditions in engineering because its factories had been subject to regulations during the war. However, wartime conditions sometimes had a negative effect, for instance where blacking-out factory windows and skylights greatly reduced ventilation.

Postwar, employers had to attract workers not only by good wages but also by cleaner and healthier conditions. Even cotton at last improved its ageing buildings. But the image of the factory as noisy and dirty remained. Conditions governing work practice were altered in wartime and the alterations lasted until 1951.

Engineering Unions

During the war, hundreds of thousands of new jobs were created in engineering, attracting the previously unemployed and the low-paid from other industries. Thus trade unionism grew rapidly in engineering, accompanied by shop steward organisation in many plants. This did not obstruct the improved dilution agreements made between the EEF and the skilled unions of the rapidly expanding engineering industries, though naturally implementation was not always smooth. The peak of hours worked in engineering was in 1942, when fitters worked an average of 54.7 hours, and their labourers 57.3 hours. The AEU had recruited skilled workers (mostly), but during the war it increasingly recruited the semi-skilled. In 1942 it admitted women, and within a year had recruited 139,000. Membership of the AEU reached over 900,00 by 1945. The TGWU also recruited actively among new men and women workers. Some unions stuck to all-male and craft traditions; the National Union of Foundry Workers, for instance, refused to admit women. The Foundry Workers Union in 1944, and the AEU in 1946, joined the CSEU but the unions retained their separate agreements and separate identities.

The transition from war to peace in engineering went smoothly, and it was here that the phrase 'made redundant' gained currency. 'Redundant' was the expression civil servants used to indicate that a factory was no longer needed for war production: it did not refer to people. Gains made by trade unions in the war were not lost but, although agreements on the employment of women and on the Joint Production Committees remained in force, they fell into disuse. Full employment seemed assured, trade union membership held up, and firms unionised during the war remained unionised. Where new machinery was installed, new agreements were made on manning. Elsewhere, old ones were restored by 1952.

Cotton Textile Unions

In 1939/40 the main limit to cotton production was a shortage of labour. The application of the Essential Works Order to the industry in 1941 caused a great deal of resentment among the workforce, since they could not then leave for better-paid jobs available in munitions in Lancashire. Nevertheless, employment in the industry fell by 42 per cent between 1937 and 1945. In 1942 the spinners were asked to work overtime. Over the year as a whole, it was calculated that even though overtime was worked for half the year, production was barely above what it would have been had no overtime been worked. The combination of increasing fatigue and machine breakdown counteracted the increase in hours.

After the war, there was still a shortage of labour in the comparatively low-wage industry. Little investment had taken place and output depended almost entirely on the number and quality of the labour force. Determined efforts were made to encourage recruitment after the Essential Works Order was lifted, but better-paid jobs elsewhere and memories of unemployment in the 1930s made people ill-disposed to believe that cotton would provide a secure future. European immigrants were recruited but there were not enough workers for all the machines.

During the spring and summer of 1948 cotton employers were subject to pressure to speed up production for exports. Borrowing something from Russian techniques, Cabinet Ministers descended on Lancashire to address large meetings of trade unionists and employers, and delegations were invited to meet the Prime Minister. Publicity of all kinds was given to the need for greater production. What, then, was the result of all these efforts? The answer was, not much: production increased a little, but only temporarily and the workers resented the continuous exhortations to work harder. Reorganisation of work practices and investment had been postponed.

Shop stewards and Joint Production Committees did not exist in cotton, although there were close local contacts between officials of unions (usually men, in a female-dominated industry) and employers' organisations. Trade union officials in cotton were rather special because the extreme complexity of the wage system meant that they had to pass highly technical exams to qualify for union office. The officials were resistant to change and although they participated with employers on the Cotton Board and the Cotton Working Party, both trying to improve the efficiency of the industry, neither had much effect. The attitudes of both employers and unions were fixed, despite the efforts of government. Cotton unions were formally committed to nationalisation of the industry but they were not very enthusiastic; perhaps because they anticipated that it would lead to changes in technology and working practices. Neither unions nor employers could face the fact that the relativities had altered, skill could no longer compensate for non-investment in newer machinery, and cheaper, standard products now had expanding markets.

Shipbuilding Unions

The expansion of shipbuilding during the war presented more difficulties than engineering as a whole. Dilution produced considerable friction and some strikes because of union opposition. Shipbuilding unions clung to restrictive practices, even though there was a shortage of skilled labour in the shipyards. To fit a porthole, for instance, a shipwright marked the position, a driller prepared a central guide hole, a caulker made the opening, a driller drilled the side holes and a caulker bolted in the light (glass) and attached the chain. Each man in turn had to be summoned by the foreman. From 1939 to 1945 the proportion of skilled men in the labour force barely changed and the number of women employed was very small. This was partly due to the necessity of using skilled men, given the equipment available, but also to the attitudes of the unions to the maintenance of their control over skilled work and their anticipation of a slump after the war. In mid–1941, a strike on the Clyde was led by the Boilermakers Society, which refused to allow the Constructional Engineering Union to work on ships. The Boilermakers struck again for sole right to use a new American machine.

Payment-by-results became increasingly important and by the mid-1950s only about 20 per cent of workers were on time rates, a great contrast to the situation before 1914. Another significant change was the displacement of some riveters by welders. Safety of workers also received some attention.

After 1945 the yards were still fully occupied and shipbuilding workers were able to negotiate from a position of strength. Wages increased almost every year, and in 1947 hours of work were reduced to 44 per week. Organisationally, too, the unions were strengthened by the addition to the (rather loose) CSEU of the Foundry Workers in 1944, and the AEU in 1946: the CSEU now represented all the shipbuilding trades. Nevertheless, the unions in the shipyards themselves still remained divided on craft lines. This increased both labour costs and the unwillingness of management to invest in new technology, such as mechanical handling of steel plates, for fear of more disputes. Higher wages and shorter hours were tipping the balance of the economics of the industry to fewer men and more mechanical aids for each ship built, but shipbuilders did not make the move, and the unions' attitudes did not help. Full order books removed any sense of urgency over change.

Iron and Steel Unions

The iron and steel industry did not change much during the war and labour relations remained cooperative, but greater control was necessary. In 1941 a special scheme to prevent labour from leaving iron and steel works was introduced, the Iron and Steel Ring Fence Scheme, which allowed workers to change jobs only within the industry. This was linked to an Essential Works Order, and had the effect of improving conditions in those sectors where collective bargaining was poorly developed. In South Wales there was a strike at the works of Richard Thomas for recognition of the staff section of the TGWU. This succeeded, but only after Bevin threatened to take over the firm if it did not concede. At times, the Ministry of Labour exerted pressure to increase the wages of the lowest paid, especially in noisy and dirty work such as drop-forging or ore-quarrying itself.

The Iron and Steel Trades Confederation, the industry's trade union, had declared itself in favour of nationalisation in 1944, but showed none of the enthusiasm for nationalisation that the miners did. There was no trouble in recruiting additional labour, and somewhat larger furnaces, a higher proportion of which were electric, helped increase output per man-year by a third, 1937–1950. Given the wages structure of the industry, this meant higher wages for many. Those workers whose wages were related to the *price* of steel had their wages structure altered to rise with the increased cost of living, a move necessary for good relations once the price was stabilised by the government. Nationalisation in 1949, theoretically operative in 1950, had barely any effect on the unions before it was ended in 1953.

Chemical Unions

During the war and postwar years the attitudes of employers in the chemical industry were forced to alter because of the increased prestige of trade unions and a social and political climate increasingly favourable to reform. ICI, hitherto somewhat paternalistic, was affected by this change; trade union membership had grown considerably, though the extent varied from plant to plant. Some were really closed shops, that is, only trade union members were employed, whereas others were barely unionised. Trade unions had always been critical of ICI's Works Councils and staff status scheme and indeed consideration was given by ICI management to possible changes here, but none came about. Although there had been recognition in some localities previously, the most important change was a national agreement in 1947 which recognised shop stewards throughout ICI.

The war years saw a great expansion in membership of the Chemical Workers' Union, which interwar had been a tiny left-wing union operating mostly in London among employees of chemists' shops. The union was unusual in that it was not allowed membership of the TUC, despite repeated attempts, until after the war.

Rayon Unions

In the rayon industry, too, trade unionism became increasingly important and managerial attitudes changed. Samuel Courtauld had argued in an influential article in 1942 that a trade unionist should be put on the boards of companies. His firm did not do this though it, too, had to recognise that the earlier high-handed attitude to the unions had to alter. Courtaulds successfully prosecuted strikers at one of its mills during the war. Gradually the firm became accustomed to negotiating with the unions; wage claims which would once have caused outrage on the firm's board were viewed in a more matter-of-fact manner.

Motor Vehicle Unions

Unions in wartime secured recognition from employers in firms previously engaged in the motor industry. Vauxhall recognised trade unions in 1941, but they remained poorly organised. Although in the USA Ford had been forced to recognise unions in 1941, in Britain Ford management held out against them until in 1944 Ford workers at Dagenham staged a strike, demanding recognition of trade unions. Under government pressure, Ford accepted the right of workers to join a union and for unions to negotiate, but only through national officers and not district or local officials, and certainly not through shop stewards. The aim was to prevent shop steward organisation within the factory. However, later in 1944 Ford was forced to recognise shop steward representation after a strike of 11,000 workers. Morris, too, was forced to recognise trade unionism. Once this was agreed, rival trade unions recruited members in all car factories. When wage restraint was abandoned in 1950, this led to difficult sets of competing claims. Shop-floor organisation was particularly strong at the car firm Standard where high wages, which the firm could afford in the postwar sellers' market, were extracted. In 1945 the firm conceded a closed shop to the unions. Furthermore, shop stewards and workers had a great deal of control over the process of production. Workers were organised in 'gangs' which set their own pace of work, led by shop stewards who had semi-managerial tasks. The firm employed a relatively small proportion of managers. The case of Standard is exceptional; workers at Ford had no such control over work, but Standard illustrates the great changes that had taken place since the 1930s.

With fluctuating markets, contending unions, and managements unused to negotiating with them, the motor vehicle industry provided a fertile ground for future and much-publicised strife.

Conclusion

Trade unions were able to exert great influence on industry and gained considerable prestige, both at national and plant level, 1939–51. Labour became a vital commodity in short supply, much wanted during the war and after it. The government had a strong incentive to ensure that workers cooperated fully in the war effort, dealing openly and directly with their organisations, and in Bevin had the right man to do it. In addition, the political commitment of the labour and trade union movements to victory over the Nazis was complete and this enhanced their acceptability. The Labour Party's 1945 landslide victory added to the confidence of the unions.

Unlike 1919–25, the postwar years from 1945 to 1951 did not see great fluctuations in demand or wildly rising wages, though the Korean War greatly affected some prices. Workers felt that they deserved something for their efforts during the war, and the continuance of controls and rationing at least provided some of the 'fair shares' they wanted. Despite newspaper headlines which made the worst of every strike — and rubbed in the irony of a Labour Prime Minister's preparedness to use troops to unload ships in a dock strike — there was considerable harmony. However, part of the price of that harmony was too often a return to old ways of work once wartime agreements about work practices ended. Unlike Germany and Japan, which had to start from new amid the ruins of their industries, and reinforced by the limitations on new building and an atmosphere of restriction, British employers accepted that return, and trade unionists remained locked to that mode of production.

Further Reading

Bullock. A. (1967) *The Life and Times of Ernest Bevin*, Vol. II, Heinemann.
Fox, A. (1985) *History and Heritage: The Social Origin of the British Industrial Relations System*, Allen and Unwin.
Inman, P. (1957) *Labour in the Munitions Industries*, Longman.

Questions

(1) What were the similarities and differences between laws affecting trade unions and their members, 1914–18, and 1939–45?
(2) There was considerable harmony between unions and governments, 1940–50. Why was this?
(3) Could unions in the cotton textile industry have done more to assist that industry's output and international competitiveness?

Section D

1951–1980

… of the two … robot welding lines for the Metro

10

British Manufacturers and the Economic Environment

The years 1951–70 were on average years of unprecedented growth and prosperity in most countries and in the world economy. There was no war between European powers, and while there was large-scale participation by US forces in Vietnam (ending in 1973), turmoil in Southeast Asia, though tragic, was localised. Russia, Eastern Europe and China did not participate much in world trade. In Africa, new nationalist governments wrestled with problems caused by internal (sometimes violent) opposition, growing populations, and rising expectations. In Central and South America, rapid economic growth, helped by oil discoveries, at times moved ahead of population growth, to give the promise and the actuality of a rising standard of living. The Middle East prospered in its oil-based wealth and Japan emerged as a major economic power. However, there were many political, military and economic anxieties.

The Suez affair of 1956/7, the Israeli-Arab War of 1967, and the deliberate actions of OPEC (Organisation of Petroleum Exporting Countries) disrupted the flow of petroleum oil. Oil, essential for land, sea and air transport for most of the world, was also the base of much of the chemical, plastic, and textile industries by the 1970s. OPEC restricted its output and organised a fourfold oil price increase in 1973 and a further threefold increase from 1979 to the beginning of 1981. Coupled with the new spending power of the oil-producing countries, these actions caused economic shocks everywhere and curtailed worldwide economic growth in the 1970s.

Had governments done nothing more than regulate tariffs and provide coinage, they would have had to consider the impact of these events and make choices. Most governments, including the British, had accepted responsibilities which took them far beyond this. Therefore, there were expectations to be satisfied. Some of the aims and choices of British governments, 1951–80, in a challenging world, are now considered.

STRONG POUND – ↓EXPORTS ↑IMPORTS

WEAK POUND – ↑EXPORTS ↓IMPORTS

Perspective: Employment, Inflation, and the Value of Sterling

Political parties in Britain in the 1950s, 1960s, and most of the 1970s, agreed that they should influence the economy to maintain full employment. Three per cent unemployment, once accepted as a rough *minimum*, became the desired *maximum*. 1951–1971, as an annual average, it had never been reached. Governments faced with *any* rise in unemployment increased demand (in order to encourage employers to take on more employees). They made hire-purchase terms (government-controlled until 1972) easier, so raising immediate consumer demand for goods, or reduced taxes or tax levels so that consumer spending power increased. Direct government expenditure and that of the nationalised industries was raised at times, increasing jobs in national and local government and their agencies (schools, hospitals, road and housebuilding, for instance) and also increasing purchases by government from the private sector which, it was hoped, would take on more people to produce the wanted goods.

If the money for such government expenditure could not be met from taxes, or by borrowing from sources which would otherwise have used it for different projects, the government could in effect create money. This too was done, ahead of extra goods and services being produced. As Keynes had anticipated, it allowed inflation to happen. If inflation showed signs of getting out of hand, thereby threatening those on pensions and fixed incomes and also threatening the sale of exports and hence the value of the pound, one remedy was to try to stop price rises and the wage increases which were passed on with price rises. In the 1950s, this was attempted by urging private industry not to increase prices, curbing price increases in nationalised industries, raising taxation and hire purchase terms to delay consumer spending, postponing investment in nationalised industries, and increasing interest rates (very cautiously) to limit credit and monetary expansion. In the 1960s and 1970s, both Labour and Conservative governments made wage and price freezes compulsory for short periods, and the Prices and Incomes Board (1964–70) and the Prices Commission (1973–77) had the power to limit or forbid specific price or wage rises.

These opposite policies of encouraging and restricting demand were known as the policies of stop-go. Their timing was often inappropriate. The state of affairs had often altered by the time the measures had effect. They could not control import prices. Nevertheless, the vehicle of the national economy continued to lurch forwards, although it did so more slowly than other comparable economies.

An additional factor was the value of the pound. The disadvantages of lowering it have been described in Section C, though an immediate effect of a fall in value might well be an improvement in exports. However, when the pound was devalued in 1967 by 14.3 per cent, wages and prices climbed upwards to reach a new level, cancelling out the advantage gained far more rapidly than after the 1949 devaluation. In the 1970s, the pound was allowed to float, or at least be flexible in value within limits. It ranged from $2.40 to $1.64 in value. This did not help the economy as the critics of the policy of keeping up its value had hoped. Indeed, in the 1970s many of the apparent links between employment, inflation, interest rates, and raising or lowering the value of sterling seemed to hold no longer; it began to be said that less inflation, not more, was the recipe for reducing unemployment. A stronger pound had been the object of government policy in the 1950s and 1960s, even if that meant raising interest rates to tempt foreign holders of sterling to keep their money in it, in spite of fears that high rates would lead to less investment in industry in Britain. When the 1979 combination of a huge increase in world oil prices and the coming into production of major British oil rigs boosted the value of the pound, this helped the money reserves but, in the absence of tariffs, encouraged imported manufactured goods, which became cheaper to buy, thereby adding to the difficulties and mounting unemployment in British manufacturing industry.

With many of the over-simple generalities of the past in tatters, economists modified their models of how the economy worked. It was apparent that a new factor had been introduced into the economy in the 1971 government decision to act on the White Paper called Competition and Credit Control. This expanded credit and put more money into the economy. More money still came from the oil profits of the Middle East, and without these two infusions there might not have been a general price rise; there might instead have been an increase in oil-related prices after 1973, but a proportionate fall in demand for many other things. Monetarist economists claimed that the key to correct and beneficial government dealings with the economy was in its control of the money supply. The purist version of monetarism held that this was best done by restricting government borrowing, a key element in the expansion of credit and money in the economy. The less pure version additionally allowed the government to use changes in the interest rate to encourage or discourage credit seeking. The idea was that, with inflation brought under control through these policies, government could withdraw from intervention in the economy.

While the Conservative governments of 1979 onwards, led by Margaret Thatcher, accepted some of this doctrine and certainly had no sympathy for any type of national plan, they nevertheless took steps to promote small businesses, improve inner cities, and so on. They denationalised industries and 'privatised' jobs previously done by public sector employees, and put emphasis on creating a climate in which manufacturing and other industries should not look to the government for support (even though at times the government gave it). The Labour Party, whose spokesman in the late 1970s had declared that it was not possible for a government to spend its way out of a recession, nevertheless held to the view that the good running of the economy was part of the responsibility of government. The new Liberal–Social Democratic Party Alliance adopted a stance of selective encouragement, a position somewhere between that of the two older parties.

Complaints, Criticisms, and Controversies

Great publicity has been given to Britain's loss of relative share of world markets in manufactured goods, to its occasional absolute drops

Figure 10.1 *Growth in Real GDP over the Periods 1964–73 and 1973–82 (constant 1975 prices)*

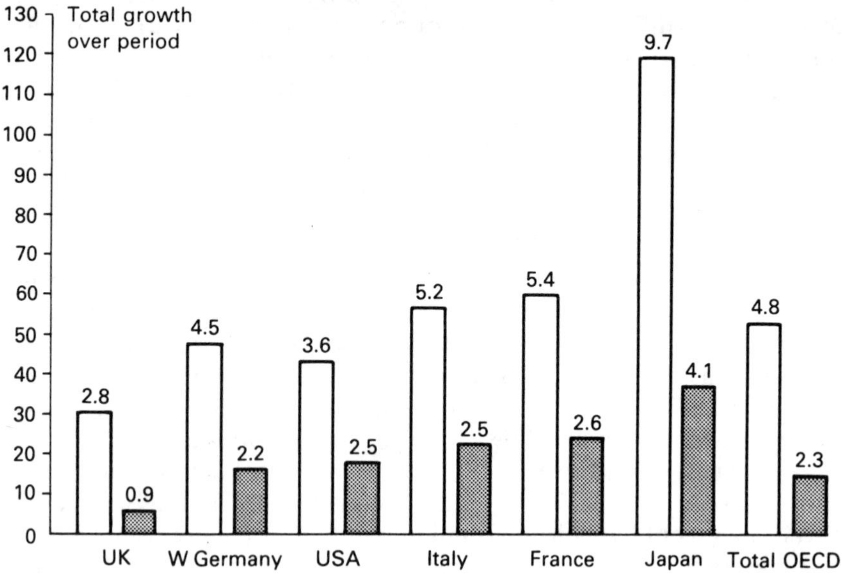

Key: 1964–73 □; 1973–82 ▒; figures in bars are % p.a. trend growth rates.

Source: NEDO (1983) *British Economic Performance*, p. 3, NEDO publication.

in manufacturing output and to the reasons for the faster relative growth and larger absolute level of the standards of living in other developed countries. The figures are not in doubt, as Figures 10.1 and 10.2 show. Much of this weak comparative performance was undoubtedly due to Britain's comparatively slow growth — and occasional falls — in manufacturing output, 1951–80.

What was in doubt were the causes and remedies for this state of affairs. Was there 'too much' employment in the government sector? Did trade unions obstruct output and increase prices by insisting on overmanning and poor working practices? Should (and could) management have invested more in new technology? Was management inefficient? Was North Sea oil, which kept up the value of sterling, the reason for the decline of the manufacturing sector? Were effective prices and incomes policies possible? Would they have helped performance in the longer run? Was there something wrong with British education? These questions should be kept in mind while reading the next sections. Some answers to them will be offered later.

Figure 10.2 *Standard of Living: Real GDP Per Head (constant 1975 $ prices — purchasing power parities basis)*

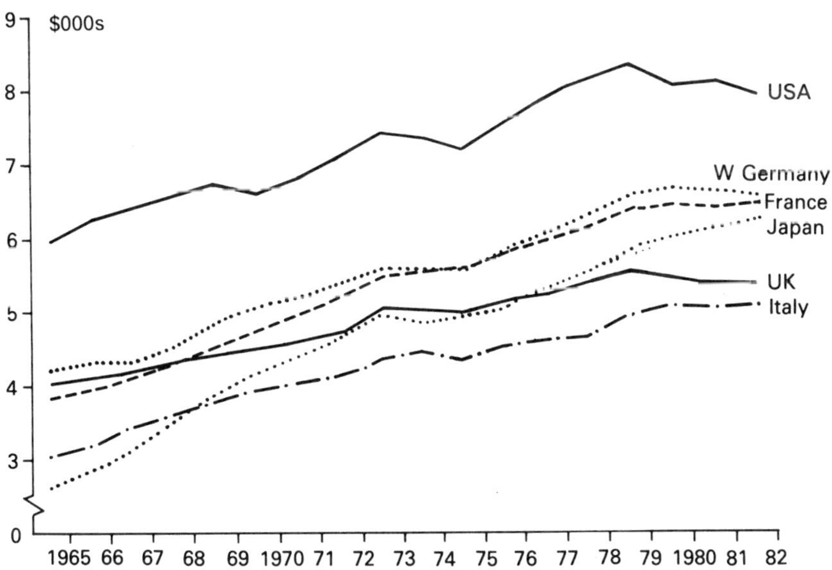

Source: NEDO (1983) *British Economic Performance*, p. 3, NEDO publication.

Features and Factors

From 1951 to 1966, output of manufacturing grew 3.2 per cent per annum on average, and then at 3 per cent per annum, 1964–1973. The amount of capital going into industry was a record average for Britain of 3.3 per cent per annum; employment in manufacturing increased by an average 0.2 per cent per annum from 1951 to 1964 (1965–6 was the peak year in absolute numbers employed in manufacturing); then began a decline of 1.6 per cent per annum, 1964–73. As these figures suggest, total factor productivity in manufacturing improved by 2 per cent per annum, 1951–64, and by 3.1 per cent per annum, 1964–73.

The combination of the 1973 oil price rise, its disruptive effect on the world's economy, Britain's own home-grown crisis of the 3-day week (see below) and entry into the European Economic Community (EEC) then caused a setback. Manufacturing output fell, 1974 and 1975, and numbers employed in manufacturing fell faster still, at 3 per cent per annum, 1974–1980. Manufacturing as a percentage of total British output declined from 35.7 per cent in 1951 to 30.1 per cent in 1973, and then to 24.5 per cent in 1980. Almost all major producing countries also experienced a temporary decline in manufacturing output, but Britain's was the most marked, and so was its decline in manufacturing employment and its rise in total unemployment, 1974–1980. Yet, in manufacturing, labour was increasingly more highly paid in the countries depicted in Figures 10.1 and 10.2 than in Britain. Increasingly also, total factor productivity, aided by a higher level of investment in technologically advanced equipment, was greater abroad than in Britain, as it had not been in previous epochs. There was a connection.

Most countries devoted a higher proportion of gross and net (i.e. total minus replacement) investment to manufacturing industry than did Britain. It could be argued that in the early 1950s much of this was still making up for wartime destruction and neglect, but capital investment continued abroad and the profit on that investment was greater than in Britain. This encouraged more investment still. At constant 1975 prices, the picture is depicted in Figure 10.3. It is against this background that features and factors between 1951 and 1980 are now described.

Size of Firms

By 1980, 491 firms in manufacturing employed 1,500 or more, often in more than one plant. In 1951, 24 per cent of the total manufacturing workforce was in plants of over 1,500; by 1980, it was 30 per cent. There was a decline in the number of middle-sized plants and a marked

Figure 10.3 *Profitability: Net Rates of Return to Fixed Capital, Non-Financial Corporations*

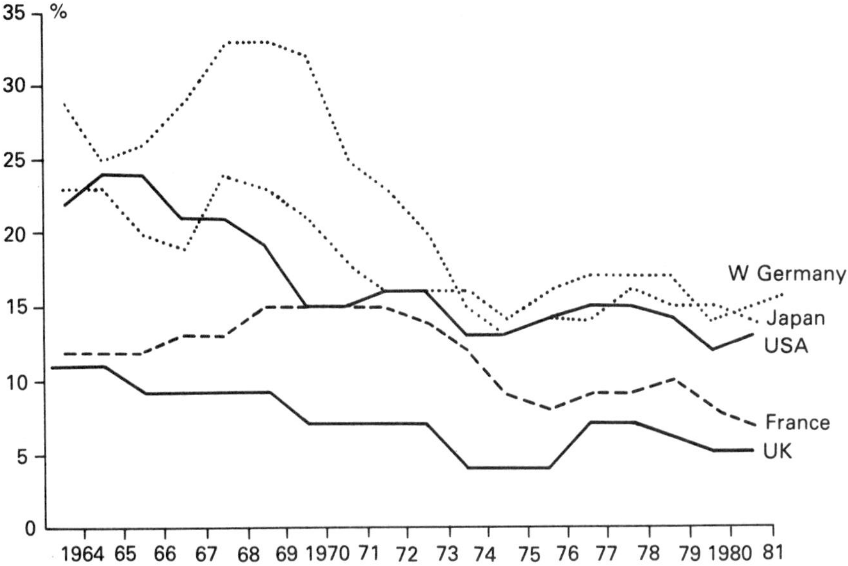

Source: NEDO (1983) *British Economic Performance*, p. 47.
Note: West German figures relate to manufacturing only.

increase in small manufacturing establishments in the 1970s. Where the total number of manufacturing plants had been about 70,000 in 1968, it was 75,000 in 1972 and 98,879 in 1980. Even before the Conservative Government of 1979 brought in no less than 108 different schemes in five years to help them, small firms, some in computer-associated industries, were on the increase. It also became apparent that bigness was no guarantee of profitability or employment.

Mergers and Concentration

Throughout the 1950s, accelerating in the 1960s, and with a peak in 1968, mergers increased. This was the main reason for an increase in concentration of manufacturing firms. Mergers supplied not only extra capacity but also ready-made customers. Bank lending was restricted by the government in the 1960s, and share prices of some major companies were low. This meant that managers were reluctant to turn to the stock exchange and unable to use banks to raise more capital. On the other hand, if shares (and profits) of a company they wished to acquire were lower still, they could 'buy it out' with a further issue of their own shares to the shareholders of that company. Merit has been claimed for the takeover bid because it puts companies on their toes. They then become more profitable, and thereby avoid a takeover. This must be good for the vigour of their enterprises, it is argued. It certainly brought larger fees to the organisers of the takeovers. At times, it made firms pay more attention to short-term profits than to long-term growth. This, carried too far, is bad for the firm and bad for the economy.

Newly-merged companies did not always achieve hoped-for profitability. A study, aptly called *Disappointing Marriage* (Meeks, G., 1977, C.U.P.), showed this to be true on average (though individual companies of course did better). One difficulty was in keeping spirits up amongst the taken-over executives (or at least, those who were kept on). Another was the avoidance of large-scale redundancies after a takeover, even though they made economic sense. Further, some mergers were formed at too high a price. Dividends had to be paid on a greatly increased capital. Yet large firms which did not acquire other firms did not grow very much, if at all, in the 1960s and 1970s; those which did, also grew by investment in new plant and new products. Where the 100 largest firms in Britain shared under 20 per cent of the market in 1914, they had 33 per cent in 1958, and about 40 per cent in 1970, and half of the 100 in 1974 had two or more products which took at least 25 per cent of the individual markets for those products.

This increase in concentration led to the Monopolies and Mergers Act of 1965 which enabled the Board of Trade to refer to the Monopolies Commission any merger in which the value of the acquired company was more than £5 million (this figure was later increased), or where a merged firm would have one third (later one quarter) of the market. 833 proposed mergers were looked at, 1965–72; only 20 were referred, and of the 20, 7 were voluntarily stopped, and 6 allowed. The Fair Trading Act of 1973 allowed the Commission to investigate uncompetitive practices in monopoly situations; by 1982, with investigations covering services as well as goods, only 40 cases had been looked at. The total effect on manufacturing was slight and so was the apparent degree of consumer protection.

More effective was the Restrictive Practices Court (1956). Unlike the Commission, which had to *prove* that mergers (and, later, monopolies) were not in the public interest, the Court *assumed* that agreements between firms about market sharing were not in the public interest, *unless* firms and industries could show that they were. Reinforced and made clearer by later laws, the legislation caused restrictive agreements to melt away, although doubtless some informal understandings remained. Resale price maintenance by individual firms was also brought into the net in 1964: exceptions were permitted in the public interest — such as minimum prices for new books on the grounds that this minimum encouraged and enabled publishers to produce more new books.

Further, a spur to competition and to consumer interest came with the scaling-down of tariffs, first within EFTA (European Free Trade Area) and then within the EEC (European Economic Community). Finally it became clear that for some products — popular-range motor cars, aircraft, and large computers — the most economic scale of production was bigger than the British national market. Effective international competition or, as managers saw it, rivalry to the point of extinction of the national industry, posed new problems in practice and in monopoly theory.

Location of Industry

As full employment continued in the 1950s, and pre-war differences in employment between regions were reduced to a fraction of 1 per cent, the government granted IDCs more readily in places outside the Development Areas. Regional aid fell from over £11 million per annum to about £3 million. New firms and expanding old ones experienced the pre-war pull of London and the Southeast, but nevertheless there seemed to be work for all.

By 1958, however, a decline in coal, cotton textiles, railway equipment and shipbuilding meant the reappearance of unemployment in the older industrial regions. This was 'structural' unemployment again, although the strengthening of national demand usually resulted in some slight improvement even in regional figures. New local measures were introduced under the 1960 Local Employment Act. In 1967, Regional Employment Premiums were paid to local employers to supplement wages for people whom they would not have employed otherwise. By 1970, added to by both Labour and Conservative governments, there was a great variety (some of it changing too fast for continuity) of Area help. The 1970 Conservative Government, in order to simplify and limit government expenditure, tried to cut back capital grants and include more tax benefits to firms siting plants in depressed districts; instead, an increase in unemployment in 1971/2 made it pass the Industry Act of 1972, which added to previous provisions. By the mid-1970s, there were Area tax concessions for factories, equipment, and land; capital grants, low-rent 'advance' factories, and the regional employment premiums all were on offer. At peak, 47 per cent of the British workforce lived in Areas designated under development schemes of one type or another, and they covered over two thirds of the

United Kingdom. Expenditure on standard grants, and selective assistance for land and factories, reached almost £489 million in 1977/8. Help had also been given to Area road building and infrastructure generally, and the clearance of derelict sites.

In direct national cost–benefit terms, much of this seemed worth doing because the cost of providing housing and an extended infrastructure in more prosperous regions would have been far greater for local and national government and for the whole economy than the cost of Area aid. Further, the multiplier effect was recognised. By the late 1970s about half a million extra people had employment in assisted Areas. If that does not seem many, it should be recalled that in 1971 total unemployment in the United Kingdom was under 700,000, 3.4 per cent of the working population.

Unemployment grew and the cost of all the schemes mounted. The Labour Government in 1976 ended the Regional Employment Premium, though it had begun a Job Training Scheme in 1975. It was thought more sensible to encourage jobs anywhere, since there was some unemployment everywhere, than to entice firms to areas which were not their first choice. Moreover, flaws became apparent in existing schemes. Grants had been given to firms which put up capital-intensive plant employing comparatively few people, so that the cost per job provided was high. Oil companies received millions which they would have spent anyway. While earlier studies showed that, in hard times, parent firms closed down older plants and kept new ones in assisted areas in production, the 1970s saw some spectacular closures of 'development' factories. Further, a decline in sales of British cars led to the transformation of the prosperous West Midlands into an area of high unemployment. The old Area definitions were inadequate.

Public money for *all* industries, 1951–80, far outweighed aid to the distressed regions. But although the 1979 Conservative Government was pledged to limit regional help, it rose, and the 1980 creation of Enterprise Zones, small areas where all but basic planning requirements would be removed, and 100 per cent capital grants made available, added to the total.

Two areas of new growth were the Strathclyde region of Scotland, where the Scottish Development Agency (established by the Labour Government in 1975 together with a Welsh counterpart) encouraged high-tech industries, and the 'sunshine belt', a ribbon of development serviced by the London–West Country M4 motorway, which owed rather more to private enterprise. In the latter, proximity to London, and an absence of reminders of a waning industrial past (which might sour both the landscape and industrial relations) outweighed the government-provided inducements of the older regions.

Policies aimed at helping local and structural employment problems made sense, but one might question the wisdom of the localised rehousing programmes of the 1960s and 1970s, especially in some northern towns and cities. That people needed better housing is unquestionable; that it was built, and often shoddily built, with capital-intensive roads, drains, schools, etc., where present and future employment prospects were grim, was to perpetuate a problem. Perhaps more thought and less localised financial control could have built basic housing where work was available, rather than renew it where it was not.

National wage bargaining was also no incentive to the relocation of industry in areas of high unemployment. The differential — some 5 or 10 per cent lower than the national average — was not a sufficient enticement. Table 10.1 illustrates the shifting pattern of regional unemployment and helps explain why any job, rather than a job in a particular place, became a priority.

Table 10.1 *Unemployment Percentage by Region, 1966–83*

	1966	*1975*	*1983 (est.)*
Scotland	2.7	5.2	15.0
North	2.5	5.9	17.3
Northwest	1.4	5.3	15.7
Wales	2.8	5.6	16.1
West Midlands	0.8	4.1	15.7
Yorkshire and Humberside	1.1	4.0	14.1
East Anglia	1.4	3.4	10.6
Southwest	1.7	4.7	11.3
East Midlands	1.0	3.6	11.7
Southeast	0.9	2.8	9.5

Source: Prest, A.R. and Coppock, D.J. (1983) *The UK Economy*, p. 229, Weidenfeld and Nicolson.

A new development was the decline in employment in manufacturing in large cities. Between 1960 and 1978, they lost a quarter of their manufacturing employment, and London lost 43 per cent. The population of inner cities also declined, but not as rapidly as this type of employment. Perhaps local planning restrictions, intended to improve the environment by separating manufacturing industry from housing, were all too effective.

Population, the Manufacturing Workforce, Employment, and Education

Population in Britain rose from 48.9 million in 1951 to 54.3 million in 1981 in spite of a slight increase in the death rate. The earlier fall in the birth rate meant that the percentage of older people in the total population grew. However, where 48 per cent of men aged 65–69 worked or wanted to work in 1951, only 16 per cent did so in 1981. The greater provision of private and government-aided pensions, the greater accumulation of goods during a lifetime of higher pay than their parents earned, the greater difficulty of older men finding employment, and perhaps the stresses endured by men of that generation during the war years 1939–45 explain this.

The birth rate rose in the later 1950s, reaching a peak in 1964. This led to an increase in numbers seeking work in due course. In the 1970s, the birth rate fell. The school-leaving age was raised to 16 in 1972/3 and, encouraged by the greater provision of places at universities and newly-created polytechnics (and greater availability of maintenance grants for the students themselves), the proportion of young people in tertiary education rose, so that some 16 per cent of the male age group 25–29 years spent at least a further three years in education by 1980. This removed many young people from the job market for a few years.

For a time in the 1950s, emigration to Australia, Canada and South Africa outweighed immigration into Britain, but it was soon more than counterbalanced by arrivals from the West Indies, India and Pakistan, and some from Ireland and southern Europe. In the long run this produced significant changes in the racial and cultural nature of the workforce, especially as immigrant families tended to be larger than those of the 'native' British. The employment effect of the newcomers was overshadowed by the influx of women, particularly married women, into employment. Where in 1951, 22 per cent of all married women were in employment, the 1981 percentage was 50 per cent. The female age group with the highest participation was 45–59 years.

The net effect was that the working population increased by 0.8 per cent per annum 1951–55, 0.5 per cent 1955–60, 0.7 per cent 1960–64, was static 1964–68, and grew 0.1 per cent 1968–73; from then to 1980 it increased by an average 0.5 per cent, or 130,000 a year. Unfortunately, this last increase coincided with harder times and made jobs harder to get.

In manufacturing in the 1960s there were often complaints that it was difficult to get workers. In 1966, to 'shake out' labour from distributive trades and miscellaneous services, a Selective Employment Tax was introduced, doubled in 1968, halved in 1971, and abolished in 1973.

Manufacturing industry nevertheless decreased its workforce, 1966–73. The difficulty, it was still said, was in finding skilled people, not sheer numbers.

The percentage of the workforce having manual skills rose slightly, 1951–1971 (it averaged about 36 per cent), but the percentage of manual workers, skilled and unskilled, in manufacturing industry as a whole fell. The absolute number of skilled people declined, due in part to the unwillingness of firms to bear the cost of training or of apprenticeship, since they had no guarantee that young people, once trained, would stay with them. In 1964 Industrial Training Boards were established; they put a small levy on all but the smallest firms in an industry, then used it to compensate firms which had appropriate training schemes. The 1973 Employment and Training Act modified the system, abolished when the assorted training schemes of the 1980s were introduced. In general, the gains in skills made during the 1939–45 war were not carried forward. People with good technical, theoretical and mathematical abilities were, however, needed and were in short supply.

The expansion of university and polytechnic education was not geared to any estimation of need for this type of education. Instead, following the Robbins Report of 1963, the inclination of school leavers was the dominant influence on subjects to be studied. In absolute figures, the number of scientific and applied science students rose (though not all science students became scientists) but was a lower proportion of all university and polytechnic graduates.

In 1973, young men under 20 were no more likely to be unemployed than the all-male average. By 1983, they were twice as likely to be unemployed as the all-male index suggested. Firstly, firms economise on labour in hard times by reducing recruitment, and this hits inexperienced young people most. Secondly, it has been calculated that a 3 per cent increase in the relative cost of employing males under 18 causes a fall of about 60,000 in their employment. Instead, firms hire 45,000 adults, more experienced, and less likely to change jobs voluntarily. Uncertainty about redundancy laws reinforced hesitations about taking on new people. At the other end of the age group, a quarter of all vacancies and a third of all manual vacancies specified an upper age limit of 50 by 1980. In general, the greater the absence of skills, the less proportionate demand there was for a worker, and youth and late middle age were barriers to new employment.

By 1983, over 40 per cent of women were part-time workers, but only 5 per cent of men. The industries — such as coalmining and shipbuilding — with the highest unemployment rates were those with fewest women employees. The increase of female employment was in the main growth sectors of the public services. Of these female employees, 55 per cent in education (teachers, secretaries, dinner ladies) worked part-time, and so did 40 per cent of professional and scientific female workers (doctors, nurses, ward helpers, laboratory assistants). It may be that the nature of the work helped create the supply and that traditional attitudes prevented men from applying for these jobs. In any case, public sector employment of this type has been cut back since 1980, and the female unemployment rate has risen without any fall in the male rate to 1986.

Capital

Capital investment in manufacturing was encouraged by governments through a variety of tax incentives. There were also grants for ailing firms and industries and grants for high-tech high-flyers. In the 1960s and 1970s, official bodies such as the Industrial and Commercial Finance Corporation could make capital available to promising firms. Small businesses received increasing capital help by the 1980s, and of course there was Area assistance.

The profitability of companies declined to 1976 (see Figure 10.3). There was then some recovery — after the average had been improved by the removal of bankrupt firms from the total on which the average was calculated. Small wonder that *profits* as a source of capital fell from almost 70 per cent in 1964 to under 60 per cent in 1980. (West Germany's comparable percentages were 72 and 74 per cent.)

Capital was more readily available from *banks* after the government opened the way in 1971. Manufacturing firms at first were very wary of using it. Between 1971 and 1973, interest rate on bank loans rose from over 5 per cent to over 15 per cent. However, since the rate of inflation was higher still, real interest rates averaged 2.4 per cent, 1974–9. By 1980 almost 33 per cent of funds for capital investment in non-financial corporations in Britain came from banks. Financial institutions claimed that there was no shortage of capital in Britain, only a shortage of enterprise in making use of it. Would-be small borrowers pointed not only to the high nominal interest rate, but also to the assets (including the family home) which they were required to put at risk as security for such loans.

By 1975, over a third of all shares in British companies were owned by pension and insurance funds. It was said that they were averse to investing in new, high-risk, but potentially high-return industries, but some created special 'venture capital' funds for such investment. Mostly, though, they were passive shareholders.

From 1951 to 1980, capital investment was a lower percentage of

GNP in the United States and in Britain than in other countries. These two countries had the most extensive stock exchange services. Was there a connection? Were profits used for dividends rather than ploughed back into the firms? Given the relatively small proportion of capital supplied by the stock exchange (11 or 12 per cent in Britain) this cannot be the whole story, but there were notable examples such as the British Motor Corporation. This contrasts with German firms, whose bankers supplied some of their capital and nursed them through periods of high investment, not asking for immediate returns, in the belief that in the long run higher profits would result. The low return on investment in Britain had some of the elements of a self-perpetuating problem.

Export of Capital and Capital Imports

A further sign that capital originating in Britain, or built up by returns on British assets, was not in short supply lies in the continued availability of long-term capital for export. There were restrictions, not to prevent long-term investment abroad, but to finance it by ploughing back profits already earned overseas, or by borrowing foreign currency. From 1965–78, about half direct investment overseas was financed from retained profts; since then, the percentage has been higher: when controls ended in 1979, the flow of capital abroad shot upwards, but subsequently fell.

In 1962, 13 per cent of British investment was in Western Europe and 23 per cent in North America. By the mid-1970s, while the North American percentage was the same, 28 per cent was in Western Europe, reflecting the changing pattern of trade and the influence of the EEC. As a percentage of GNP, capital outflow declined, and after 1960 the return to Britain of interest, profits, and dividends on investments held abroad was greater each year than that year's long-term capital outflow. Even so, those receipts were less than 1 per cent of GNP. Far more important to the economy were the 'invisible' returns on services abroad and, greater still, returns on manufactured exports.

By 1962, about 55 per cent of newly added net assets of manufacturing in Britain were foreign-owned, though this fell to 40 per cent by 1985. Conversely, British multinationals (including service sector companies) continued to invest abroad: in 1983 their foreign assets were four times those of foreign multinationals in Britain. A subsidiary of a multinational in Britain at least provided local employment and had multiplier effects: the alternative might have been an increase in imports from the parent firm, or a diversion of its investment to some other country.

Short-term capital movements usually (but not in 1976 or 1978) were greater inwards to Britain than outwards. After the 1979 oil price rise, the value of sterling strengthened; this attracted short-term foreign capital, offsetting about 40 per cent of long-term outflows.

The Domestic Market

Downwards fluctuations in earnings, unemployment and employment were minimal for 20 years, 1951–71, compared with 1919–39. It was a period of high domestic demand. Thereafter, demand fell, 1973–5, as unemployment grew and world trade fell. In 1979–81, however, while unemployment rose, demand and consumption fell less.

A rise in manufactured imports began in the 1960s and early 1970s, *predating* the higher value of the pound due to the flow of North Sea oil, and given extra impetus in 1979 by that oil and by that higher pound, which meant a lower comparative price for imports. The share of the final market for manufactured imports in Britain increased eightfold between 1955 and 1983 until, in 1983, Britain imported more manufactured goods than it exported. This was a shock.

Of course, when GATT, joining EFTA and then the EEC, lowered British tariffs, it was likely that more British consumers would buy some foreign goods, even when they cost more — but especially when they cost less. Further, as the standard of living in Germany and France overtook that in Britain, British manufacturers had a smaller home base from which to operate.

Export Markets

Worry over the reduced share of British manufactured exports in world trade pre-dated concern over imports. After the enormous postwar effort, the British percentage fell as shown in Table 10.2. Some of this

Table 10.2 *British Percentage of World Trade in Manufactured Goods (%)*

1950	25.4
1954	20.5
1959	17.7
1964	14.2
1969	11.2
1974	8.8
1979	9.1
1984	7.6

Source: *Report from the Select Committee on Overseas Trade* (1985) House of Lords, Session 1984–85, Vol. I, p. 23, HMSO.

is explicable not just by the increased output and exports of other countries, but also by the absolute fall in British manufacturing output between 1973 and 1976, and the much steeper fall between 1979 and 1981, though these in themselves need explaining. (Nevertheless, the volume of British exports then steadied, and was as high in 1984 as in 1980.) Some exports were of goods which were also imported. Differences in design and performance accounted for this type of trade, similar descriptions masking differing degrees of quality and sophistication.

British exports prices fell by 13 per cent, 1967–1974 (that is, after devaluation) relative to prices of products of other countries, yet British exports' share of world trade fell by 26 per cent. Subsequently, although British prices rose relative to those of other countries, this did not reduce British exports correspondingly. One study concluded that price accounted for only half of the loss of export share between 1956 and 1976, and one fifth between 1970 and 1976. The rest, on this analysis, was due to inferior design (springing from inadequate R & D), late deliveries, and poor service. Price, however, undoubtedly had an important impact from 1979, when oil drove up the pound's value.

A different explanation is that British goods were not being produced in sufficient quantity to take up the potential percentage of world trade. Since for so much of the time employment in Britain was virtually full, the argument goes, output must have been near the limit of capacity. This might have some truth — but only if it is accepted that technology could not have become more productive. But it could.

German and Japanese export industries grew on technological development and investment and should have been the pattern for Britain. The direction of manufactured exports also changed — towards the EEC, North America, and the oil-rich Middle East, and away from the old Commonwealth countries.

The Role of Sterling and the Balance of Payments

The value of sterling and the balance of payments have been discussed on pp. 180–1. In order to obtain money from the IMF in 1967 and 1976, when sterling and the balance of payments were both precarious, the government had to accept IMF's conditions, including limitations on public expenditure.

Government and the Economy

The role of governments in increasing and decreasing demand has been discussed. The body of government rules and regulations affecting manufacturing industries, dwindling in the 1950s, increased again in the 1960s and 1970s. The idea of deregulation and of decreasing taxation to provide incentives first found vigorous support in the United States in the 1960s: 'supply -side' economics, it was hopefully called. In Britain, in addition to demand management, attempts were made at times to curtail government employment and expenditure on the grounds that it was wrong for it to occupy more than one third of the total economy. The arguments were obscured by a political argument over whether nationalised industry was part of a non-productive government enterprise, or the cornerstone of an advanced economy on the Swedish rather than the US model. There were also arguments over 'transfer payments' — that is, taxes and national insurance contributions which took purchasing power from one sector of the population (higher wage and salary earners) and gave it to another (pensioners, mothers of children, unemployed people). Should these payments be regarded as a part of government expenditure?

After 1951 the income tax threshold came at a lower *percentage* of average earnings but those average earnings in real terms increased. By European standards, most British income tax as a proportion of income was not particularly high and 'social charges', as a proportion of GNP spent on social services, was about average for Western Europe in the 1960s, and 1970s. It is impossible to judge just how far, if at all, income tax was a disincentive to investment or enterprise; perhaps the greatest deterrent effect of taxation came through estate duties on personally-owned businesses. There was also criticism of corporation tax, levied on company profits since 1945. This left less profit to use for expansion; however, it may also have had the effect of causing companies to try harder to earn more.

Governments gave help to exporting firms through the Export Credit Guarantee scheme, and through promotional displays and commercial advice abroad. The government took Britain into EFTA in 1959. Britain agreed to reduce tariffs reciprocally with the other members — Austria, Denmark, Norway, Portugal, Sweden and Switzerland. This certainly increased Britain's trade with them though, amongst other things, it caused the collapse of the British paperboard industry. The other major European group, the EEC, had grown out of the original European Coal and Steel Community. Its six signatories — Belgium, France, Germany, Holland, Luxembourg and Italy — agreed in 1958 gradually to reduce trade barriers between themselves. However, the EEC's wider objectives at the time seemed incompatible with Britain's Commonwealth ties. As the latter loosened, and EEC growth forged ahead, Britain reapplied to join, and was admitted to full membership in January 1973, leaving EFTA.

The sum total of benefits and burdens which EEC membership would entail could not be worked out in advance. Government economists provided estimates which differed tenfold from each other. Manufacturing industry found that the entry of EEC goods into Britain grew faster than the reverse trade. Full monetary union was not achieved.

In 1962, a Conservative Government established the National Economic Development Council (NEDC) to analyse problems in the British economy, identify obstacles to faster growth, and point out to government and industry what should be done to put things right. Smaller Economic Development Councils (little Neddies) were established for particular industries. Building on this, the Labour Government of 1964–70 organised a Statement of Intent on Productivity, Prices and Incomes, signed by the Government, the TUC, the FBI, and other bodies. It emphasised cooperation in the interests of full employment, faster growth and rising real incomes. A National Plan was drawn up — a plan only in the sense of indicating what, under certain circumstances, ought to be possible — namely, 4 per cent annual growth. Alas, circumstances (particularly sterling reserves) altered, the target was scaled down, and then there was more caution about what was called 'indicative planning'. The future role of government in the economy remained the topic of much discussion.

Conclusion

We can make some tentative answers to some of the questions raised in the Complaints, Criticisms and Controversies section. Reducing the government 'share' of the economy is not easy in a time of recession; not only are transfer payments to unemployed people to be made, but governments, if only for electoral purposes, spend to try to improve the performance of the economy or at least curtail unemployment. Declining employment in the government sector has not been followed by declining unemployment in the private sector. North Sea oil, with the higher value it gave sterling, especially after 1979, did make exports more difficult and imports easier, but the relative decline, and the occasional fall in absolute British manufacturing output, predated 1979. The difficulties in controlling an economy, or even its money supply, when that economy was affected so much by world trade, was apparent. Other questions raised in that section are best answered after reading the next two chapters.

Further Reading

Foreman-Peck, J. (1983) *A History of the World Economy: International Economic Relations Since 1850*, Wheatsheaf Books.
Pollard, S. (1982) *The Wasting of the British Economy: British Economic Policy 1945 to the Present*, Croom Helm.
Pollard, S. (1983) *The Development of the British Economy*, Edward Arnold.
Report from the Select Committee on Overseas Trade (1985), House of Lords, Session 1984–85, Vol. I, HMSO.

Questions

(1) How do you account for the comparatively steady growth of British manufacturing and the British economy, 1950–70?
(2) Was it sensible to have regional or area policies?
(3) Was 'stop/go' a good way to run the economy?
(4) What were the consequences of inflation for British manufacturing industry? Could devaluation at a given moment or a floating pound have solved them?

11

Old Industries, New Industries, and Managerial Choices

While government actions affected firms because they affected the economy at large, managers after 1951 were freer to make their own choices about supplies, outputs, markets and the workforce. Government focus was largely on the behaviour of the economy as a whole.

There were exceptions. The PIB, the Monopolies Commission and 'Little Neddies' affected individual firms, and governments intervened in the old staples and in car manufacture, and even in electrical engineering because of falling sales, profits, and employment in those industries. Managers welcomed their financial and other help. Most of the firms were traditionally exporters, contributing much to a favourable balance of payments, and this strengthened government interest and involvement.

In 1963 Harold Wilson, who became Labour Prime Minister in 1964, said 'We are redefining and we are restating our socialism in terms of the scientific revolution . . . the Britain that is going to be forged in the white heat of the revolution will be no place for restrictive practices or outdated methods on either side of industry'. There were two measures to help the forging. The first was the creation of a Ministry of Technology, which from 1965 onwards gave public money to firms designing and producing computers, machine tools, electronics and communications, and subsequently aided aircraft production and design. The second was the establishment of the Industrial Reorganisation Corporation (IRC) in 1966. It had £150 million to encourage mergers which would promote efficiency and put vigorous management in charge of new entities. This kind of 'interference' was welcomed at least by managers in the dominating firm. Not all IRC's creations were financial successes, but without it more of British manufacturing industry would have gone out of business.

The Conservative Government of 1970–74 abolished the IRC but, in spite of statements about not helping 'lame ducks', gave money to ailing firms. The next Labour Government in 1975 set up the National

Enterprise Board (NEB), not primarily to rescue firms, but to 'back winners' by buying government shares in them. It was occupied largely by the first of these roles. The Conservatives in 1979 determined to distance government from industry, but this had had little effect by 1980. It was against this changing background that managers' choices were made.

A feature of British firms was that, until the late 1970s, output did not fluctuate with the ups and downs of trade as much as did output abroad. Various explanations were suggested: a plausible one was that British employers and managers were loath to sack workers in difficult times since, from 1940, they had experienced a shortage of skilled and reliable workers in a situation of virtually full employment, and so kept them at work. Undoubtedly this tendency was reinforced by fear of employee ill will should dismissed people be needed again, and by some uncertainty over new redundancy laws. But it meant additional dismissals when recession came in 1979.

The percentage of GDP contributed by manufacturing was 36.5 per cent in 1960, 32.0 per cent in 1973 and 24.3 per cent in 1982, accompanied by a 30 per cent decline in employment in manufacturing, 1970–80. In absolute figures, British manufacturing output fell, 1973–76, recovered somewhat, and then fell even more, 1979–81. Other sectors of the economy grew, but not fast enough to absorb the unemployed. What happened in manufacturing was clearly of great importance to the individual employee, and to the economy as a whole.

It becomes less and less realistic to separate industries into 'old' and 'new'. Yesterday's 'new' industries become today's old ones. Old industries and products — textiles or machine tools — adopt new inputs and techniques. Therefore, while covering much of the same field, the grouping of industries is somewhat different in this chapter from that in Chapters 2, 5, and 8.

OLD AND NEW INDUSTRIES

The Textile Industry

Cotton

Cotton goods from low-wage India, Pakistan and Hong Kong (some of the latter actually from Japan) entered Britain free of tariffs in the 1950s. Competition from low-wage countries was perhaps expected; the improved quality of these imports was not. The Cotton Industry Act 1959, under which firms were compensated by the government for scrapping excess machinery (£11.3 million) and workers who were made redundant were paid a lump sum, was the result. The government also paid 25 per cent of the cost of re-equipping firms that decided to stay in business. Some 48 per cent of spindles and 38 per cent of looms were destroyed. The least efficient firms left the industry, and some remaining ones re-equipped with labour-saving machinery, so productivity increased significantly. However, the basic problems — a structure which distanced firms from the final customer, and too much old machinery — remained. Cotton merchants minimised their risks by reducing stock, and yet did not seem able to pick up market trends and send correct guidelines back to manufacturers. Subsequently, total annual imports from Hong Kong, India and Pakistan were limited to a quota of about 20 per cent of British consumption, and in 1966 quotas were imposed on imports of all but European, North American, and developed Commonwealth countries. Other developed countries evolved similar policies.

Courtaulds sold rayon yarn to cotton manufacturers, and so had an interest in their survival; ICI had built up textile expertise in nylon and terylene which it believed it could transfer, and it was interested in mixed fibre (including cotton) yarn and cloth. Later, Courtaulds wanted cotton-type spinning for its acrylic fibres. So Courtaulds and ICI jointly financed English Sewing Cotton's takeover of the ailing cotton firm of Tootal. ICI then bought shares in a number of important cotton firms and provided loans for mergers and re-equipment. Eventually, in 1970, ICI bought Viyella and Carrington and Dewhurst and merged them to form Carrington Viyella (later Vantona Viyella).

Courtaulds bought two of the largest cotton spinning firms, the Lancashire Cotton Corporation and Fine Spinners and Doublers, thereby controlling about one third of cotton spinning. Weaving firms were usually much smaller than spinning and often not suitable for modernisation. Therefore Courtaulds built new weaving plants in development areas, using government grants and imported machinery. It also acquired finishing firms, clothing firms and retail shops, and became probably the largest textile concern in the world.

This looked hopeful: ICI and Courtaulds had textile experience and management and marketing skills. They were helped by retained quotas on imports of cotton and cotton-mixed textiles and a tariff imposed in 1972. The new large firms shut down old, inefficient cotton mills and reorganised work (Viyella reduced Combined English Mills' mills and spindles by half, employees by two fifths, output by one tenth, but increased output per spindle by 63 per cent and per employee by 53 per cent). Most firms used a mixture of cotton and other fibres; Viyella and ICI pioneered the polyester/cotton sheet.

But in spite of increased domestic consumption of cotton and cotton-mix textiles, the UK manufacturers' share of this domestic market fell from 54 to 30 per cent. Additional imports of the 1970s were largely from *higher* wage cost countries — the EEC, the USA and Japan. The main reasons were the quality, design and service backup offered, rather than a price advantage over comparable British goods, though at times they had this too. Curiously, British exports of mixed cotton and man-made fibre woven fabric increased while their output and domestic consumption declined. Perhaps the newer big companies were too big to change course swiftly to meet changing British taste; most certainly the fashions of the late 1970s for cotton corduroy and denim were met by imports. Some of the great decline in employment in cotton (from just under 200,000 in 1961 to 54,000 in 1980) was due to increased productivity reducing the need for workpeople, but at least half seems to have been due to closures and diminished domestic demand. The import of ready-made clothing from India also diminished British demand for cotton cloth. Worldwide, textile output grew much faster than consumption, and Britain was one of the principal sufferers. The firms themselves, especially Courtaulds, felt that governments had not appreciated that the industry was no longer a low-wage, low-tech industry, and it needed tariff or quota protection for a longer time to recoup its high capital outlay.

Woollen and Worsted

In the 1950s, these industries seemed to be performing quite well, recovering from the 1951/2 downturn in demand for all textiles.

From the mid-1960s onwards, the story was not so happy. Of the 991 firms in the industry in 1963, only 331 remained in 1980. While a little of this fall was due to merger — Courtaulds acquired some well-known firms — much was due to bankruptcy. 163,000 employees in 1965 fell to 50,000 in 1980, and output fell from 260 million square metres to 118 million. In constant values, British output fell by 26 per cent from 1965–1980, French output was static, German increased by 1 per cent, and Italian grew by 34 per cent. While upmarket British woollens sold well, the 'cheap and cheerful' fashions of the 1960s and later told against wide sales of quality goods in Britain.

A 1969 Neddy report had recommended help, but it was 1973 before Stage I of the Wool Textile Scheme, with loans, grants, and reduced interest rates, was implemented. £16.2 million was taken up. The outcome was a boost to modernisation and some elimination of excess capacity, but no mergers or change in industrial structure (though some occurred outside the scheme). Further, the payment of a

Temporary Employment Subsidy prolonged the existence of uneconomic firms.

Stage II (1976) was more selective and successful. It caused restructuring, some *within* firms — for instance by rebuilding and concentrating production on one site. There was new help for small firms with assets of £50,000 or less and a grant of 50 per cent of the cost of external consultants to advise on improving their performance. £6.1 million of public aid was given under Stage II. Twenty-two per cent of firms for a time achieved the desired 100 per cent increase in productivity, due largely to newer, faster, machinery — over 80 per cent of it foreign-made. A Code of Practice covering employees helped reduce suspicion of the schemes; very little use was made of its retraining clauses, but a good deal of its redundancy compensation. The new machines were easier to operate (it took 4 weeks instead of 2 years to train fly-spinners), but took up to 6 hours to set up for a given type and texture of cloth. This, coupled with the 1979–1980 recession, had an unforeseen consequence. Old machines took only about half an hour to set up; thus when orders were for short runs of cloth, they were cheaper to set up and run. Therefore old machines which should have been scrapped under the scheme were allowed to continue.

The firms themselves thought good management the key to future profitability. They put fashion towards the end of the list of desirable qualities, thereby underestimating its impact, as imports of upmarket Italian cloth and finished garments showed. The average British man and woman, however, was more concerned with price than quality: hence the manufacture of polyester/wool suitings, and the imports of Comecon cloth.

Scottish firms, especially those allied with SEET (Scottish, English and European Textile group) and with well-known brand names, did better than the average Yorkshire firm. Ten principal buyers for retail outlets took 60 per cent of the output of British woollen and worsted industries in 1980; it was on their orders and judgement that the firms depended, though exports to a diversity of countries continued. By 1980, manufacturers were encouraged by signs of a return to more formal wear for men, which of course raised demand for suits and overcoats, a new interest in natural fabrics for clothing among women purchasers, and an appreciation of the non-static qualities of woollen imports. Italy, France and Germany liked good-quality British woollens and worsted, but not cheaper British output.

Man-Made Fibres and Textiles

The new oil-based fibres of the 1950s and 1960s — nylon, polyesters,

Table 11.1 *BNS Nylon Output, 1947–1961 (m. lbs)*

1947	0.935	1951	11.70	1955	32.1	1959	54.2
1948	1.31	1952	12.714	1956	34.7	1960	71.2
1949	3.554	1953	14.427	1957	44.4	1961	80.7
1950	9.319	1954	21.914	1958	45.5		

Source: Coleman, D.C. (1980) *A History of Courtaulds*, Vol. 3, p. 198, The Clarendon Press. See also Tables 14, 17, pp. 79, 179.

and acrylics — reduced rayon output. By 1950, the BNS nylon plant, the first nylon plant in Europe, was at full production and capacity was thereafter increased, as Table 11.1 shows. BNS was taken over completely by ICI in 1964, and ICI shed its large shareholding in Courtaulds. Thereafter, Courtaulds went on to develop and manufacture its own nylon.

Though the fibre was invented by scientists working for Calico Printers Association, ICI spent about £10 million in developing Terylene, its brand of polyester, demonstrating that production of new man-made fibres needed large firms with adequate capital. Terylene, like nylon, became a commercial success. However, ICI lost £3.7 million on 'Ardil' between 1951 and 1957, due to its poor wet strength and its relatively high price. ICI subsequently had success with 'Crimplene'.

Courtaulds in 1952 took the first steps towards a multidivisional structure, which should have improved its organisation. It diversified into paints and packaging. Late into acrylic fibres, it developed a process that did not infringe patents. It also developed a good dyeing process for the new fibre which it named 'Courtelle', launched with great success in 1958–62, some compensation for the falling profits of viscose rayon. Courtaulds bought the smaller British Celanese in 1957, when its profits too were falling, and one way or another, all other viscose rayon producers disappeared by 1960. Courtaulds' diversification was not a success, and competition from EFTA countries reduced Courtaulds' profits on staple viscose to two thirds of its mid-1960s level by 1971. By the late 1970s two thirds of *all* fibres and textiles used in Britain were synthetic. The growth of output is shown in Table 11.2. Foreign chemical firms such as Dupont, Hoechst, and Monsanto established plants in Britain to manufacture fibres.

Following the rise in the price of oil, its basic feedstock, in 1973 the synthetics industry was in crisis. Slow world growth and the rise in costs led to huge overcapacity worldwide. Courtaulds and ICI shut plant. In 1978 the European fibre makers tried to set up a cartel to control prices but EEC restrictions stopped them. They did however cut capacity in

Table 11.2 *UK Production of the Three Principal Synthetic Fibres, 1963–70 (m. lbs)*

	1963	*1966*	*1968*	*1970*
Nylon	135	185	275	300
Polyester	55	105	150	170
Acrylics	10	15	60	66

Source: Moncrieff, R.W. (1975) *Man-made Fibres*, pp. 989–990, Newnes-Butterworths.

1978 and 1982. The recession after 1979 further hit the industry. Long-term orders from suppliers of bedlinen and clothing to chain stores became essential to survival, and further contractions of synthetic and natural fibre textile firms followed.

The Iron and Steel Industries

Until 1953, nominally still nationalised, iron and steel firms were run by their pre-nationalised managers. The Iron and Steel Act (1953) set up two bodies, one to sell off the companies to the private sector (86 per cent were sold by 1957) and the other, the Iron and Steel Board, to coordinate the activities of the now separate companies. The Conservatives justified this intervention by the tariff protection which the government gave the industry, and the need to continue allocations because of the shortfall in supply. Labour renationalised steel in 1967, forming the British Steel Corporation (BSC) from the major firms. The 1970 Conservative Government did not denationalise it.

During the 1970s, a number of major engineering and similar privately owned firms extended backwards into the production of special qualities of steel, primarily for their own use. The Conservative Government of 1979 was committed to denationalisation of iron and steel, but when the recession of 1980 seriously affected these private companies, BSC acquired and financed some of their plants, thus *adding* to the nationalised firms.

The 1950 British capacity of 16 million tons per annum proved insufficient; steel had to be imported: the revised plan was for 22.5 million. Restrictions and limitations on exports continued until 1957, and then some sort of balance between home demand and supply was reached. British steel was not expensive by world standards in the 1960s: labour costs per man-hour were relatively low although, since its furnaces were smaller and equipment less mechanised, more man-hours per ton of crude steel produced were required.

In 1955 and 1960, when there was a strong demand for products made of steel, British manufacturers bought steel abroad to meet delivery dates (rather than wait). Some continued doing this. Nevertheless, British firms bought mostly British steel, and from 1950 to 1966 output doubled. At the same time, world output quadrupled, led by an astonishing 34 per cent per annum increase in Japanese output, 1952–1969. When there were dips in home demand, British steel usually could be exported, and the increased capacity and the greater capital outlay seemed justified. In South Wales and at Ravenscraig in Scotland, new strip mills were built: the decision to build two mills rather than one was based in part on the wish to spread employment.

The British Steel Corporation produced an even more ambitious programme, doubling capital outlay (at constant prices) between 1970 and 1976. But the 1970s saw a further great increase in world production and in the meantime the demand from British shipbuilders shrank as that industry shrank; and the car industry, like many others, was using lighter, tougher steel, and its car sales fell, so it too ordered less steel. Government attempts to curb public expenditure, especially from 1977 onwards, meant less steel needed for construction. Add to this the world recession of the late 1970s and early 1980s, and it is no surprise that from 1978 onwards steel mills, usually but not always the oldest and smallest, closed down. The industry was at a crisis. The cost of new investment could not be paid for from the quantity of steel sold. British prices were now higher than world prices, partly because of the cost of production in under-used plant. A long steel strike (see next chapter) added to this. In 1980, output was only 40 per cent of capacity. This improved to 15.3 million tons, i.e. about 50 per cent of capacity in 1981. It was clear, nevertheless, that demand was not going to recover to former levels. Cheapest steel came from South Korea which combined high investment, large-scale plant, low labour costs, and indirect US subsidy. Steel employment fell from 450,000 in 1970 to 270,000 in 1980 (175,000 in 1983). As only the more productive plants survived, productivity improved, helped by new manning levels and better management practices.

Technical improvements had been made. The amount of fuel needed for each ton of steel was reduced; the Redcar sinter plant (plant using crumbled iron ore which had been artificially made into manageable lumps) was hailed as a model of efficient production. The technique of blowing oxygen onto molten ore, rather than forcing gases through it, was eventually adopted by all major works. Nevertheless, if the iron and steel content of manufactured and semi-manufactured goods imported into Britain was added to the imports of iron and steel, and this viewed against the quantity of similar exports, it is likely that

Britain by the mid-1980s was a net importer of steel. Yet the British steel industry survived, and was still about tenth in the world. It was probably not sensible to strive for volume of output, nor to make every kind of steel wanted for every British company. The postponement of closure of some plants in the late 1970s, understandable in political and human terms, meant that when they did close, world recession, and not only the difficulties of the British economy, made it hard for ex-steel men to find new jobs. The BSC, with public funds, attracted new light industries to Corby, the ex-steel town. Many older men did not find work in these new ventures, but their wives, sons and daughters were enabled to find employment in their own community.

With the adverse trading conditions of the late seventies, 'production quotas' for EEC member countries were imposed, but these did not greatly affect Britain, because of its previous closures. Sensibly, there is now far less inclination to use steel output and sales as a measure of Britain's industrial strength and weakness. The danger is that its economic possibilities might be undervalued.

Shipbuilding

Full order books continued throughout the 1950s, producing over 1.3 million tons (gross) of new launchings a year, but profits were not used to modernise the yards. In the early 1960s output declined; average annual launchings were just over 1 million tons and it was 1975 before 1.3 million were reached again. Nevertheless, this was steadier output than the 1920s and 1930s.

Frequent and prolonged labour disputes, often caused by inter-union disagreements, led to adverse publicity, and strikes were no encouragement to future orders. In 1950, 38 per cent of total international sales of ships were from British yards; by 1956, less than 14 per cent, although if British domestic sales were added, 20.7 per cent of world output was still built in Britain. In 1973, Britain's 1 million tons was only 3.2 per cent of all world launchings.

First Japan and then Taiwan, South Korea, Brazil, Poland, and Yugoslavia built up shipbuilding industries using their own new and relatively cheap steel output; in those countries labour disputes were not allowed, and industrial wages were lower than British (though Japanese wages later rose). The Geddes Committee in 1966 listed out-of-date equipment, poor sites, and relatively high costs of labour and raw materials as contributory causes of lower orders for Britain. However, other European countries, such as West Germany and Sweden, with higher wages than in Britain, in 1972 built 70 per cent or more of their merchant fleets in their own yards, while British shipowners bought from abroad more than 75 per cent of all new tonnage registered in Britain.

Traditionally, British shipowners were directors of British shipbuilders and their ships were built in British yards. But these links were personal and informal and if shipowners saw an advantage in ordering abroad, they did so. The type of ship required was changing. Large passenger liners were less wanted, as mounting crew costs and cheaper air travel made them less competitive — and these were ships in which British yards were well experienced, if not well equipped. Shipowners wanted oil tankers and bulk ore carriers ten and even twenty times the pre-war size, or ships which could carry either cargo. Japan's new shipyards specialised in large tankers, using a 'block construction' assembly line method which did not need traditional shipbuilding crafts, and benefited from spacious green-field sites, and from a flow of standardised orders. The hemmed-in sites in Britain were not readily adaptable — nor were working methods. Between 1950 and 1975, 90 per cent of the increase in tonnage of the British merchant fleet was in oil tankers, the overwhelming proportion built in foreign yards. South Korea got indirect US financial help, because of its defence burdens, and used some of it for its shipbuilding industry.

The traditional general cargo vessel was only 8 per cent of total world output in 1975. Specialised sophisticated vessels — container ships, roll-on-roll-off ferries and liquified gas and chemical carriers — were not often built in Britain because of the yards' technological backwardness: an absence of larger cranes and heavy gear had at last caused an unbridgeable price differential and an intolerably slow construction time. The £68 million spent under Geddes' recommendations went largely into consolidating shipyard management, in the hope (usually dashed) that it would be better. The Shipbuilding Industry Board (1967) paid out £151 million to the industry 1967–1971, but only a third went to capital equipment. Sit-in strikes followed attempts to close two yards: they were kept open, though with fewer employed. £1,000 million was lent by the government in 1972 to British shipowners to order from British yards. This, too late, offset a foolish scheme which had given an investment grant of 20 to 25 per cent of the cost of a new vessel, wherever built, to British shipowners. Foreigners had hastily registered as British companies to get money from the scheme and they, and British firms, had then ordered from abroad.

From the mid-1970s, orders for oil rigs for North Sea drilling gave employment to some yards, but others, particularly after the bankruptcy of the foreign Maritime Fruit Carriers, which had placed orders in three large British yards, had little work. In 1977, the Shipbuilders

Industry Act, in what looked like a final rescue, nationalised 27 shipbuilding companies; in 1978, the new British Shipbuilders had 86,000 employees and sales of £584 million, on which £108 million was lost. In 1980, total British output was 431,000 tonnes, and imports supplied almost 50 per cent of British home demand. The British merchant fleet was shrinking rapidly, however. Exports were helped by a variety of government financial measures.

It is unlikely that the British shipbuilding industry could have grown as fast as those of developing countries, but with the investment levels of European countries, and a break with past traditions, a share akin to the latters' (though these too were hit in the 1980s) would have been possible. The few British firms, such as Austin and Pickersgill (A & P), which correctly identified a gap in the market for bulk carriers and for standard ships within a given size range, showed what could be done.

Naval orders did not rescue the yards (except for submarines at Vickers at Barrow-in-Furness) because the naval programme was restricted; the collapse of Maritime Fruit Carriers may be seen as an unavoidable blow, or yet another example of British misjudgement. Increasingly, shipowners avoided British registration, since the standards for manning and total remuneration of British crews were higher than those of other flags. A further link in the old chain of British shipbuilding, ownership, registration, and crew was thus broken.

The Engineering Industry

In the 1960s and 1970s there was great concern about the declining exports of the UK machine tool industry and its inability to supply the domestic market in the desired quantity or quality. Machine-tool firms were unable to innovate because of their low level of R & D. The costs of development were too great for individual firms and so about £100 million was provided, 1966–1978, principally by the Labour Governments of 1964–70 and 1974–79, for R & D.

Although this had some success, it did not result in the establishment of a vigorous British machine-tool sector. In the 1970s it was in absolute decline as the figures in Table 11.3 show. British machine tools, 13 per cent of all engineering exports in 1965, stood at 6 per cent in 1978. Furthermore, machine tools imported into the UK were more technically sophisticated than UK exports. One of the most important developments in machine-tool technology was that of numerically controlled (NC) machine tools, of which computer numerical control (CNC) was a further development. Only 2.8 per cent of British

Table 11.3 *Index of Output of the UK Machine Tool Industry, 1972–1980 (1975=100)*

1972	*1973*	*1974*	*1975*	*1976*	*1977*	*1978*	*1979*	*1980*
86.6	97	92.4	106.8	88.9	82	80.6	82.2	84

Source: *Machine Tools Statistics* (1982).
Note: The figures refer to the first quarter of each year.

machine tool output in 1966, NC tools represented 14.5 per cent in 1980 and with these, instructions for operation were fed in using magnetic or punched tape or cards. The machines then worked automatically, and very accurately, repeating work exactly; they could be reprogrammed to perform other sequences. Particularly suitable for intricate work on a relatively small scale, they were developed by the US Air Force and, to a lesser extent, by Ferranti and EMI in Britain. In the 1960s Plessey acquired licences for American NC systems. With a loan from the IRC in 1969 it formed Plessey Numerical Controls and took over Ferranti's NC interests and Airmec–AEI.

Government helped two other firms (unsuccessfully) and through the NEB also rescued the leading British machine tool maker Alfred Herbert, nationalising it in 1974 and re-funding it, but to no avail. These failures suggest that the problems were more deep-seated than a lack of finance. The lack of quality and quantity of design and of professional engineering and skilled workers was blamed, but there also seems to have been some poor market assessment. Subsequently, some small firms, producing CNC tools, won some promising orders, and more British firms used these larger tools.

The pace of technical change in textile machinery greatly increased, to handle new types of synthetic yarn, and developments in engineering and electronics altered the design and sophistication of the equipment. Most British manufacturers did not keep up with these changes and lost their leadership in world markets, though they still exported 75 per cent of their output. Their traditional home market shrank. In Britain, the machinery wanted was of a type in which Britain had no long tradition. A recent study has found that British firms which paid attention to R & D were successful; the poor technical quality of the machines of other firms was the principal source of their decline. Compared internationally, British firms contained low levels of graduate scientists, and the study also criticised inefficient management. Decline was not simply a matter of price, though undoubtedly the 1979 overvaluation of the pound had some adverse effect. Figure 11.1 tells the sad story.

Figure 11.1 *World Trade in Textile Machinery, 1954–75*

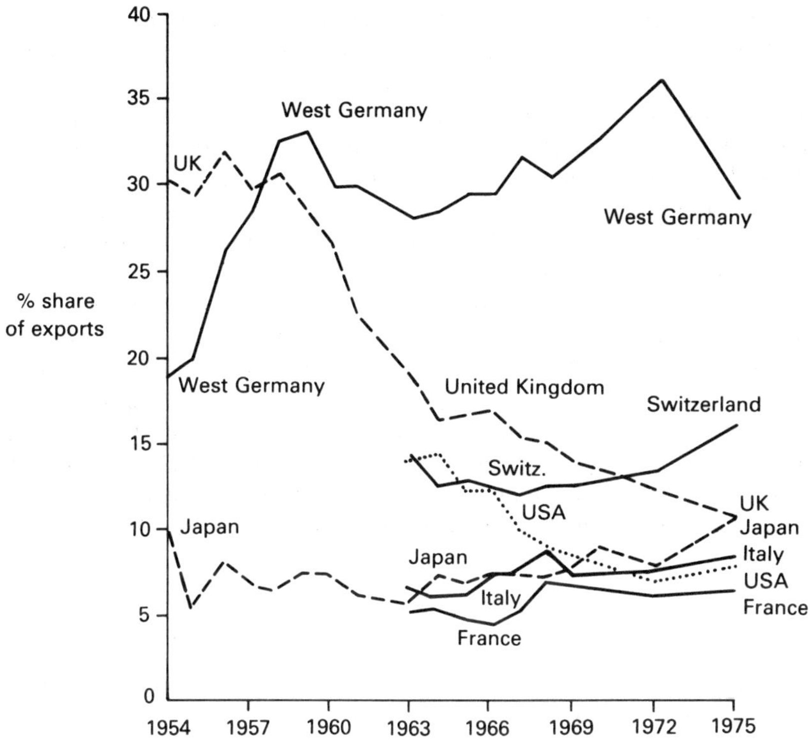

Source: Rothwell, Roy (1980) 'Innovation in textile machinery' in K. Pavitt (ed.) *Technical Innovation and British Economic Performance*, p.129, Macmillan.

One third of US typewriters produced in 1958 was electric; in Britain, only 6 per cent. Britain's most successful maker, the Imperial Typewriter Company, was slow to change from mechanical to electrical machines. In the early 1960s, under competition from imports, Imperial introduced modern production techniques. In 1966, it was taken over by Litton Industries, an American multinational, which transferred production from the USA and brought in new products, including an electric machine. However, because of poor management and low labour skills, productivity was low, so Litton closed its British plant in 1975 and transferred production to Germany. In Britain, three small plants remained American- or German-owned, working to foreign designs and producing light non-electric machines largely for export. Nevertheless, the exports of British-built machines was, in the late 1970s, worth between £20 and £30 million per annum. US machines were threatened by the Japanese and Taiwanese.

Pre-war, domestic Singer sewing machines were occasionally fitted with an electric motor. Postwar models were virtually identical. Italian designers produced machines which might have been less sturdy, but were electric, and able to do a variety of stitches with minimum adjustment. This, and the readier availability of ready-made clothing and soft furnishings, undermined Singer in Britain and its export trade. Its pioneering Glasgow factory closed amid much recrimination and the British market was supplied by imports — including imports of newer Singer models made abroad.

Fewer adults rode bicycles for getting to work, but more children and teenagers owned one. Bicycle firms amalgamated. Until the 1970s, imported bicycles were only a (rather specialised) fraction of new sales; then, changes in fashion, the high value of sterling and questionable sales strategy of British firms led to imports rising to 840,000 bikes in 1983 out of an average 1.9 million per annum sold in Britain. Imports fell after safety rules changed, but the British industry also had difficulty in exports, as low-wage Taiwan exported to the USA.

Motor Vehicles

When steel allocations conditional on exports of vehicles ended in 1953, car production for the domestic market rose three times as fast as for exports, and demand for cars grew three times as fast as incomes. Tariffs of 20 to 30 per cent on foreign cars protected the British industry. In any case, many foreign producers did not find it worthwhile to produce right-hand-drive cars, and left-hand-drive vehicles were at a disadvantage on British roads. All these factors provided a wonderful domestic market for British car manufacturers. New car registrations rose from 138,400 in 1951 to 820,000 in 1960 and to over 1 million per annum from 1963 onwards; and in 1965, 94.8 per cent of new cars registered were British-made.

This sounds encouraging, and indeed in many ways it was. People were prepared to wait months for delivery of their choice of car. However, when delays were lengthened by the non-arrival of parts or strikes on the assembly line, this tried the patience of customers and so did a certain lack of reliability in assembly or even design. Nevertheless, until about 1972, in which year 23.5 per cent of all new registrations were of imported cars, concern about the British industry was largely concern about what had been BMC and became BLMC when Leyland Motors, with government and IRC encouragement, took it over in 1968. Ford, the other large producer, had its ups and downs, but its Cortina and Escort were best sellers. BLMC's winners were the Mini and the Austin Morris 1100/1300, and they were older types, with no obvious successors. Newer models had defects or drawbacks and the firm, renamed British Leyland, then BL, then Austin Rover, had no money for new investment or adequate R & D. The Government rescued it in 1975 with £2,000 million spread over a number of years — a big sum by British standards, but not by those of the European or the growing Japanese car industry.

In the 1950s and 1960s, it was reckoned that a sales volume of at least 100,000 of a given model a year was necessary to justify the expense of designing and tooling up for a new popular car. The models mentioned met these criteria, but others did not. In some years, had there been no strikes, BLMC would not have been able to sell all the cars produced. British customers were becoming choosier, and at the same time, apart from fleets of company-owned cars (7 per cent of registrations in 1964 and 40 per cent in 1975), liked individuality. Sales in the USA became harder because US firms built smaller cars (for a time) and German popular cars were vigorously marketed there. By the late 1970s, it was reckoned that annual sales of 500,000 of a given model were the optimum, given their increased sophistication and higher specification. British firms could not approach this quantity, but the US-owned Ford and Vauxhall 'rationalised' manufacture of engines and car bodies and models between countries and factories, to reap some of the new economies of scale.

Entering the EEC meant a scaling-down of tariffs, and the 1979 boost to the value of sterling was the final help to the influx of foreign cars — 996,000 compared with 750,000 British registrations in 1979. However, new developments in computer-aided design and computer-controlled machine tools cut down the costs of changing models (though as yet they can do little for development cost), so there may yet be a future for the comparatively small scale output of British cars. Newest production lines in Britain are as automated as those anywhere. The luxury makes of Jaguar and Daimler (for a time, in financial straits, rolled into BLMC) are produced less labour-intensively, while even Rolls Royce puts less stress on being 'hand-built'.

British motor bikes, however, continued to be hand-built. The makers were not worried by light-weight Italian motor bikes and scooters, or by the Japanese machines which subsequently appeared. But then Japan, following on the success of the film 'Easy Rider', mass-produced heavy bikes, though styled for leisure rather than for everyday transport, which was the British tradition. British firms had neither the finance nor the know-how to mass-produce and, with

Japanese sales worldwide running at over 1 million per annum for cheaper and reliable bikes, British production ceased completely with the failure of Meriden, a gallant but doomed cooperative venture.

Leyland Motors acquired a number of truck and bus-making companies and, when it became part of BLMC and then BL, the buses, trucks, vans and Land Rovers of that company at times formed about one third of the value of total output. Ford trucks and lightweight Bedford vans and trucks were successful in the 1970s, and Ford and General Motors, which owned Bedford, each made Britain the centre for computerised engineering and design for all their commercial development outside the USA.

The use of heavier and heavier goods vehicles with more powerful engines accompanied the development of European motorways. These vehicles were not at first mass-produced, and it was possible for British producers to match the new demand, if necessary extending the range of components which they bought in. Indeed, they did make sturdy trucks tailored to the individual customer. But the German firm of Daimler-Benz computerised and rationalised design and production of major components, buying in some to its specification, and automating assembly. This of course brought down the cost of production, provided that a sufficient number could be sold. The US giants and their British subsidiaries themselves bought in components and did not produce these vehicles on the same scale. The steep oil price increases of 1979 curbed the demand for heavy vehicles. By 1985 there was estimated to be 40 per cent surplus capacity worldwide for their manufacture. Perhaps it was as well that Britain was not more committed to their production than it was.

These oil prices, and the accompanying recession, caused setbacks to all vehicle production. The British industry, additionally affected by the higher pound and weak past performance, suffered more than most, and so did its West Midlands component suppliers.

Aircraft

R & D costs in aircraft are very large; the more aircraft of a given type produced, the smaller the proportion of R & D carried by each aircraft. Also, in aircraft production (and in the production of other highly complex products) costs decrease as production increases because workers and managers become familiarised with the product: the learning-by-doing effect. The bigger the market, the lower the cost can be.

Because British military aircraft had shorter production runs than US ones, they were consistently more expensive, not helped by government's policy of ordering a number of aircraft to fulfil any one function. Between 1955 and 1964 sixteen types of military aircraft were produced: the average production of each was only 168. The 1957 Defence White Paper argued for curtailing military aircraft and for greater reliance on nuclear weapons carried by missiles. This was never fully implemented; nevertheless, between 1957 and 1965, several big development and production contracts for supersonic bombers and fighters were cancelled, and US Phantom jets were imported. One cancelled project was the TSR2, a skilful but ultimately unsuccessful attempt to reduce costs by combining a number of functions in one aircraft.

Aircraft firms, with government pressure, amalgamated into two groups, the British Aircraft Corporation and an expanded Hawker Siddeley, in 1959/60. After the mid-1960s the development of front-line military aircraft and their engines for the RAF was undertaken in collaboration with European countries. The Jaguar (in service in 1973) was Anglo-French, and the Tornado, under development since 1968 (in service in 1981) was built by a European consortium (Panavia). The engines were developed and built by another European consortium (Turbo-Union). The Harrier (although produced in some versions in collaboration with the USA) and the Hawk Trainer were all-British, however. Collaboration put financial pressure on the two aircraft firms, and in 1977 they were nationalised as British Aerospace. By 1981 the firm was denationalised, its financial future uncertain.

The case for government support of a civil aircraft industry was much weaker than for military aircraft; yet, with the exception of nuclear power, the most heavily subsidised and protected sector of the British civil manufacturing industry was civil aircraft. In part, the motive was to keep together R & D teams, which could be switched to military production.

From 1945, British governments provided funds in return for a levy on sales of British aircraft, but only the Viscount recouped their outlay. They also purchased civil aircraft for military use. In the most spectacular technical success but economic failure, Concorde, the British and French governments paid 100 per cent of development costs.

In the late fifties, the decline in military orders and the success of the Viscount prompted several firms to finance their own civil aircraft projects, but these did not sell well. Governments decided to support 50 per cent of development costs of a number of civil aircraft projects; between 1960/61 and 1973/74 £747.7 million (at 1970 prices) was spent on them, including the Rolls-Royce RB211 engine (see below).

In civil aircraft too there was collaboration with Europe though, except for Concorde, Britain played a smaller and more hesitant role. In 1979 the nationalised British Aerospace joined the Airbus Industries consortium, Europe's only builder of large civil aircraft, for joint projects. Governments became more wary of supporting even small-scale civil ventures, but short-range civil aircraft such as the BAe 146 were later produced with government backing.

By 1951 all new British aircraft (except small civil ones) were jets. Rolls-Royce's Avon turbojet powered both civil and military aircraft, and the firm then made a series of highly successful engines. Bristol was the only other large-scale producer. When airframe firms were merged under government pressure in the late 1950s, engine firms and divisions also merged. Rolls-Royce then became the only British maker. Its new design, the turbofan RB211, was much larger and quieter than anything the company had built before and incorporated a radically new material for the turbine blades — carbon fibre. Unfortunately, the innovation was not a success and, in spite of having 70 per cent of its estimated finance requirements met by the government, the firm was on the edge of bankruptcy. This was a great shock, because the company was regarded as the finest example of high-quality British engineering. The Conservative Government rescued the firm in 1971, the carbon fibre was ditched, the engine redesigned, and it proved a great technical and eventually financial success. Rolls-Royce, still in public ownership in the 1980s, installed CNC and robotic equipment. It could therefore expand production rapidly, as it could not do in the past. It said that it would never again develop a new engine entirely on its own.

Electrical Engineering

The Ministry of Defence was a major customer of this industry, 1951–80, and the Post Office, the Central Electricity Generating Board, the Scottish Electricity Boards and British Rail were also customers. In the 1960s, competition greatly increased from imports and in the export market. Helped by the IRC, AEI and English Electric, both in trouble, were merged with GEC, in the hope that the latter's strong management would restore their fortunes. Plant closures, untypical of the mergers of the day, and very unpopular, followed. GEC had 206,000 UK employees in 1970, and 145,000 in 1981. Growth was slow and, though there was rationalisation of production, the firm did not become a single entity. Nevertheless, it became highly profitable. One reason for the slow growth was the decline in orders for power stations. Demand for electricity did not grow as fast as predicted, and so fewer new power stations were needed. There was a special difficulty with nuclear power stations.

Nuclear or atomic power was a spin-off of Britain's successful determination to have an independent nuclear bomb. Research for both military and civil uses of nuclear energy was done mainly by the publicly-funded United Kingdom Atomic Energy Authority, and in the 1950s Britain was in the forefront of nuclear power use. Four consortia to build nuclear power stations and their equipment were formed in 1955, and a fifth in 1957. The first generation of these power stations, the Magnox series, was relatively successful, and two were built abroad, but the next series, the Advanced Gas Cooled Reactors, could not compete worldwide with the US (Westinghouse) Pressurised Water Reactors. Great site delays (experienced also in building conventional power stations) and engineering problems reduced the consortia to one, led by GEC, by 1972. It was decided to drop the AGRs and build PWRs, but planning inquiries and accidents in US PWRs delayed decisions — and in any case, since demand was lower than anticipated, there was no longer the same urgency to build. Successive governments paid attention to the argument that it was necessary to go ahead with nuclear power stations to keep design teams together; nevertheless, no PWR had planning clearance, even by late 1986.

The expansion of electricity supplies to 1970 brought a steady demand for cable of all types; new high tension grids brought new technical problems which were successfully overcome. Telephone ownership grew, adding to cable demand, though the development of optical fibres promised to reduce demand for traditional cable for long routes. Likewise, satellite communications lessened demand for under-sea cable, but here, too, optical fibres, well understood in Britain, promised an alternative. Hopes for cable demand for cable television, however, were misplaced.

As real incomes grew, so did ownership of electrical goods. Even before Britain entered the EEC, Italian washing machines and refrigerators made inroads on the British market; after EEC entry, France and Germany expanded sales in Britain even though some of their products were dearer, backed by extensive up-market advertising and promises of reliability. GEC acquired a number of firms, reviving some and closing others. Smaller independent British firms such as Lec survived to make good quality but not lavish products. After a brief flourish, British dishwashing machines disappeared entirely. By 1982, 95 per cent of all British homes had vacuum cleaners, 80 per cent had washing machines, and 68 per cent refrigerators. A decline in the rate of expansion of ownership was therefore to be expected; the market was fiercely competitive.

Electronics

By 1964, 87 per cent of all British homes had television sets, all black and white, almost all British-made, and made to the unique 405 line standard. When colour television began in 1967, and 625 lines was adopted as standard, big firms stepped up colour-receptive sets, but not fast enough. Japanese sets, numerous and reliable, at first were assembled in Britain, but then increasingly, made there. Two British firms tried to produce sets in collaboration with Japanese firms, but withdrew. Then British firms made sets with Japanese technology under licence. Of these, Thorn EMI's sets were the most successful. By 1980, Britain was a net importer of TV sets. No other country used as many video sets as an adjunct to TV as did Britain: all were imported.

The first British computer, developed during the 1939–45 war as a code-breaker, could not store programmes. Storage in all modern computers was developed in British universities, the National Physical Laboratory, and the Admiralty. Ninety per cent of the 220 computers in use in Britain in 1960 were British-designed and built. However, British firms were overshadowed by the giant US International Business Machines (IBM) Company. Its capital resources, larger, richer home base, its volume of production and range of products and after-sales service and help with initial programming, won world markets. The government, through the IRC, encouraged a merger of British firms to form International Computers (ICL), which survived as a subsidiary of Standard Telephone and Cable, itself US-owned, and has found a niche for itself in the British and some export markets.

Computer control of production technology grew considerably in the 1970s, first installed in the chemical and oil industries in the 1960s. The combination of use of computers in robots, machine tools and design is known as a flexible manufacturing system, and here British firms have shown an inventiveness in design and programming of certain kinds of system. Computers used with electronic telecommunications form information technology systems, of the kind found in banks and building societies. Most of the installations have been made by US firms (sometimes in the nimble and cheaper Far East), but British Telecom supplied most of the telecommunications linking branches to headquarters. In the early 1970s, the Post Office, of which British Telecom was then a part, paid for the entire development of 'System X' telecommunications produced by GEC, STC (though it later withdrew) and Plessey.

Transistors, which replaced valves in electronics, and integrated circuits and microprocessors (the chip) were invented in the USA, but production increasingly shifted to Japan and to equally nimble but less-expensive countries. Large British firms also made standard chips, but then restricted output to specialist production, often for defence. In 1977, foreign-owned subsidiaries of multinationals, mostly from the USA, and often located in Scotland, produced 85 per cent of the standard chips made in Britain, utilising and adding to previous R & D. The Government in 1979 financed INMOS, to start in the USA, to learn about chip manufacture, then return to Britain. An argument about whether it should be situated in prosperous Bristol or in Cardiff, with its higher unemployment but less good communications, showed the difficulties governments and managements alike experienced — that is, how to choose between something which would benefit an industry economically in the longer run, and something else which, while socially desirable, might in the end damage the industry irretrievably and cause a loss to total national wealth and earnings. Table 11.4 gives a picture of output and employment in the industry.

Table 11.4 *Estimates of Output (£m. – current) and Employment (000s) of the British Information Technology Industry, 1970–1981*

	Computers and peripheral equipment		*Software and systems*		*Telecommunications*		*Office equipment and systems*	
Year	*Output*	*Employment*	*Output*	*Employment*	*Output*	*Employment*	*Output*	*Employment*
1970	249	51	—	—	249	86	178	35
1971	238	50	69	15	316	84	—	—
1972	252	50	80	15	350	85	—	—
1973	342	46	104	17	395	88	—	—
1974	396	45	129	19	407	87	—	—
1975	465	43	164	20	494	87	253	30
1976	564	42	221	21	531	74	—	—
1977	647	44	265	21	520	66	—	—
1978	741	43	332	22	588	68	—	—
1979	959	45	430	24	691	66	—	—
1980	1,183	44	536	27	941	68	415	24
1981	1,173	42	612	26	1,233	65	—	

Source: National Economic Development Council (1982) *Policy for the UK IT Industry*, Appendix VI, Tables 6, 7, 8, 9, NEDC.

The Chemical Industry

In the 1950s, the British chemical industry expanded, but not as fast as the emerging German and Japanese industries, or the long-established US one. Investment and productivity were not as great, proportionately. In the 1960s, the output of British chemicals grew twice as fast as British manufacturing output as a whole; ICI was the largest British producer, and the fifth largest chemical company in the world.

Coal and molasses were replaced as the basic feedstock for raw materials by petroleum, and oil companies like Shell, BP (which took over Distillers' chemical operations) and Esso became sizeable producers in their own right, and also provided inputs for other firms and industries — paints, glass, soap, and paper. Polythene, polystyrene and polyvinyl chloride (PVC) were in demand, and in the 1950s and 1960s their use became more widespread as petroleum, used in their production, became cheaper in real terms.

The processes used for their manufacture were housed in large plants, becoming more and more efficient through advances in chemistry and in chemical engineering and construction. New big plants had great economies of scale, and fewer workers per unit of output. Table 11.5 shows the growth of output. In 1957 Britain was the third largest producer of plastics in the non-communist world; by 1967 it had declined to tenth. More and more countries, including Comecon ones, built their own plant for domestic and export production.

Table 11.5 *UK Production and Consumption of Plastic Materials, 1951–1968 ('000 tonnes)*

	1951	*1953*	*1956*	*1959*	*1965*	*1968*	*Growth (% p.a.)*
Thermosetting*	130	110	158	235	312	330	6.0
Thermoplastic**	65	100	155	423	630	855	16.0
Total	195	210	313	658	942	1,185	12.0
Consumption	160	182	236	544	825	1,094	12.0

Source: National Economic and Development Organisation (1972) *The Plastics Industry and its Prospects*, Table 3.2, HMSO.

Notes: *Thermosetting, e.g. Bakelite, polyurethane.
**Thermoplastic, principally polythene, polystyrene and PVC.

From 1962 to 1965, ICI built plants which made ammonia from oil. Ammonia produced the nitrates which were used for fertiliser for Britain's intensive farming. EEC payments were a later incentive to its use, which benefited other companies such as Fisons, as well as ICI. Subsequently, the ICI plant was converted to use natural gas as its feedstock, under a favourable contract with British Gas. ICI experienced some difficulties at the end of the 1960s and in the early 1970s because of its enthusiasm for technically exciting products which nevertheless had limited market possibilities. However, these hiccups were as nothing compared to the severe shocks which the chemical industry experienced worldwide when the price of oil was so drastically raised twice in the 1970s. There was already worldwide overcapacity in some chemicals and their products, and the vast increases meant a fall in sales and a rethink of product ranges. However, on the whole the British chemical industry came through the crises reasonably well.

ICI had already taken steps to improve its labour productivity (see next chapter) and its ethos moved from that of a somewhat protected imperial company to that of a modern multinational. It increased the number and importance of its subsidiaries abroad, and increased its exports first to Europe and then to the USA. Its British workforce fell from its 1969 peak of 145,000 to 61,800 in 1983, by when it had more sales abroad than in Britain. It switched more and more of its attention to value-added speciality chemicals, which formed 40 per cent of its sales in 1980 and 50 per cent in 1984. At a time when the chemical giants of the USA lost sales and profits (partly to overseas plants which they had helped to create) and new competitors appeared in Canada, Mexico, and Saudi Arabia, ICI's expansion and profitability were commendable.

Scientific and medical research led to new pharmaceuticals. Growing quantities of antibiotics, vitamins, hormones (including 'the Pill'), tranquilisers, heart drugs were produced. Distillers built up a substantial pharmaceutical business, including the ill-fated thalidomide morning-sickness pill, but sold out to the US Eli Lilly Company in 1962.

By the mid-1970s the largest pharmaceutical firms in Britain were Glaxo, Beecham, the Wellcome Foundation, ICI, Fisons, and Boots. All were British-owned, but some 60 per cent of pharmaceuticals used by the National Health Service were imported, or made by foreign firms. However, although the Monopolies Commission refused to allow Beecham to acquire Glaxo and thereby create a larger research base, individual British firms produced and marketed a succession of excellent and profitable discoveries. Production abroad, to get under tariff or other barriers or comply with requirements of foreign governments, was undertaken by British firms. Profits on individual drugs may appear large, but have to cover the long period of R & D and time spent on research which may not yield marketable products. Firms which analyse the marketed drugs, and then make similar products, can sell them more cheaply, and they do not have to pay for the meticulous work which precedes marketing by reputable British companies.

The chemical industry remained capital-intensive, accounting in 1970 for 5.3 per cent of total British employment and 17 per cent of capital expenditure. It contributed much, directly and indirectly, to the British balance of payments, and after its uncertain nineteenth-century start, its R & D was an example to the rest of British industry.

Manufacturing and Research

The necessity of developing weapons systems and defence equipment during the war left a legacy of higher spending on R & D to postwar Britain. The National Research Development Corporation was established and funded by the government under the Development of Inventions Act of 1948; it gave small-scale yet crucial help to new projects and new products. The sum total of expenditure on R & D was, until the late 1960s, higher in Britain, as a percentage of GDP, than in all major industrial countries except the USA, as Table 11.6 shows.

Table 11.6 *Gross Expenditures on Research & Development (GERD) as a percentage of GDP (A), and the Proportion Devoted to Defence (B)*

	1964		*1972*		*1978*		*1983*	
	A	B	A	B	A	B	A	B
UK	2.32	33.2	2.13	26.2	2.20	28.2	2.26	29.2
USA	3.14	35.4	2.58	29.4	2.37	25.3	2.70	28.1
France	1.84	23.4	1.86	20.7	1.77	19.8	2.14	21.5
W. Germany	1.41	11.3	2.20	5.4	2.24	5.8	2.58	4.3
Japan	1.47	0.5	1.83	0.5	1.93	0.4	2.47	0.3

Source: Kaldor, M., Sharp, M. and Walker, W. (1986) 'Industrial competitiveness and Britain's defence', *Lloyds Bank Review*, October, No. 162, p.36.

Since there were few doubts about the quality of the research or the researchers (the 'brain drain' to the USA showed how eagerly they were sought after), why did this expenditure not have a better effect on British industry? Some answers, including the absence of professional and production engineers and of adequate quality control, have been outlined in this chapter. In design and development, Britain had only one fifth of the number of professional engineers, and in the manufacture of gears, only one ninth of the number which Germany employed. It has also been seen that manufacturers, in self-perpetuating uncertainty about markets, were reluctant to invest.

There are other features of the R & D expenditure itself which need examining. Table 11.6 refers to *all* R & D. In manufacturing, from 1967 to 1975, as a percentage of manufacturing output, British R & D funded by non-government sources *fell* from 3.3 to 2.7 per cent, that of Germany *rose* from 2.2 to 2.7 per cent, and that of the USA from 3.3 to 4.2 per cent. Then British government funding of R & D fell absolutely and relatively, research foundation support grew slightly, and expenditure by firms remained at about the same percentage of manufacturing output. To be more innovative in products and their design, firms should have spent more on R & D in British manufacturing, or, Japanese-style, have concentrated more of existing expenditure on a narrower range of products.

There is another aspect of R & D, illustrated in Table 11.6. A higher percentage of R & D in Britain went into defence than in any country (save the USA) in the Western world. This reflected the relative size of British defence expenditure, and also the desire, for security reasons, and because of trust in British-designed weaponry and defence equipment, to make and buy British. Economic reasons were secondary, though often exports of British weaponry and defence items were hoped for. In fact, for better or worse, Britain's share of world trade in arms fell from 7.4 to 4.3 per cent between 1963 and 1983. British equipment, tailormade to British requirements, was sometimes too complex for substantial export sales.

Further, it is calculated (roughly) that about 30 per cent of Britain's scientists and engineers of recent years have been engaged in defence-related projects — a disproportionate figure, since defence-equipment expenditure by the British government purchased only 6.3 per cent of manufacturing output in 1974, and 12.3 per cent in 1984. The high concentration of talent was brought about by the needs of defence-related aerospace and electronics R & D, and it had fewer spin-offs into civilian industry and exports than anticipated. The NEB did its best to nurture firms with sound R & D potential and, in spite of failures like Alfred Herbert, succeeded with others such as Ferranti. More than money is needed for successful R & D, yet sufficient money is essential to achievement.

Conclusion

There are no simple answers to the questions of why British manufacturing industry suffered such setbacks as this chapter has outlined, why the British standard of living fell below that of competitive economies, and why unemployment grew faster in Britain than elsewhere. Nevertheless, it is apparent that the answers to these questions are very interrelated, and that the wellbeing of the economy depended heavily on the wellbeing of manufacturing industry. The effect of North Sea oil on the exchange rate undoubtedly made it harder to export and easier to import from 1979. But it was the last straw to some failing firms, not the first blow, and it does not explain why more expensive foreign products were preferred, even in Britain. The points about R & D in the section preceding this should also be kept in mind when reflecting on complaints, criticisms and controversies.

It looks as though management, immersed in shortrun problems (more of which will be outlined in the next chapter), lost sight of changes needed for longer-term prosperity. Further, for many firms and industries, markets could almost be taken for granted for much of the 1950s and 1960s and this bred inattention to their needs. The devaluation of 1967 gave little long-term help, since prices and wages adjusted upwards quite swiftly. The 1970s provided an assortment of shocks, so that simply keeping production going became an achievement. As the gap widened between what could be achieved with existing technology and organisation and attitudes of both employer and employee, and what could be achieved with newer technologies and approaches, too much of British industry was left on the wrong side of that gap, finding it more and more difficult to make the leap across.

Further Reading

Blackaby, F.T. (ed.) (1979) *Deindustrialisation*, Heinemann.
Pollard, S. (1982) *The Wasting of the British Economy*, Croom Helm.
Pollard, S. (1983) *The Development of the British Economy*, Edward Arnold.
Williams, Karl, Williams, John and Thomas, Dennis (1983) *Why Are the British Bad at Manufacturing?* Routledge and Kegan Paul.

Advanced Reading

Kaldor, Mary, Sharp, Margaret and Walker, William (1986) 'Industrial competitiveness and Britain's defence', *Lloyds Bank Review*, October, No. 162, pp. 31–49.

Questions

(1) What during this period were the main problems of the textile industries and what were the best solutions?
(2) Why was the iron and steel industry in crisis by the late 1970s?
(3) Why did the aircraft industry not keep up its wartime impetus?
(4) What were the strengths and weaknesses of British involvement in design, production, and marketing of computers or machine tools?
(5) What are the major strengths and weaknesses of British industries in this period?

12

Employees, Trade Unions, and British Manufacturing Industry

Because governments from 1951 to 1980 accepted responsibility for the good performance of the economy and were in almost continuous anxiety about the balance of payments, exports, inflation, or a combination of all three, they became involved in larger disputes or strikes which affected their aims. That involvement was highlighted by the blaze of publicity which covered strikes, and newspaper headlines which asked, 'Is Britain Governable?'

As Table 12.1 shows, strikes increased from 1969 onwards. About 80 per cent of them were apparently about pay, but pay was sometimes the concrete matter around which other discontents gathered. That unions should use their strength to extract wage increases when demand and employment were high was understandable. That they should use it, and use it successfully in the short run as demand and employment shrank, was less expected and less welcome. It was also used to postpone redundancy, or innovations which threatened redundancy. Between 70 and 90 per cent of all strikes were unofficial, that is, not called by full-time union officials, but by convenors and shop stewards who did not go through the negotiating processes laid down in their union's rule book. The rule books, however, did not forbid strikes outside these rules.

The number of shop stewards grew from 90,000 in 1961 to 175,000 in 1968, and 250,000 in 1978, of whom 118,000 were in manual occupations in manufacturing. The amount of time they spent on union duties varied greatly, and so did employer tolerance of their meetings and discussions during working hours. Some shop stewards received special training from their unions. While this enabled them to bargain more effectively, and to put their immediate concerns into a wider context, it might also have sharpened their reaction to grievance. Other possible reasons for growth in the number of disputes were awareness of wage increases abroad (where, however, they were often tied to increases in labour productivity), the first compulsory wage freeze of 1966 which directed attention to great inequalities of income in society, often not catching much attention in the routine of the workplace or the

Table 12.1 *Employment-Related Indices, 1960–83*

	(1) Change in retail prices (%)	(2) Change in average weekly earnings (%)	(3) Unemployment (%)	(4) Unfilled vacancies (%)	(5) Days lost in industrial disputes (m.)	(6) Change in import prices (%)
1960	1	6	1.7	0.9	3	0
1961	4	7	1.5	0.9	3	–2
1962	4	3	2.0	0.6	6	–1
1963	2	4	2.3	0.6	2	4
1964	3	9	1.7	0.9	2	3
1965	5	8	1.4	1.0	3	1
1966	4	7	1.5	1.1	2	1
1967	2	4	2.3	0.8	3	0
1968	5	8	2.4	0.8	5	12
1969	5	8	2.4	0.9	7	4
1970	6	12	2.6	0.8	11	4
1971	9	11	3.4	0.6	14	5
1972	7	13	3.7	0.6	24	5
1973	9	14	2.7	1.4	7	28
1974	16	18	2.6	1.3	15	46
1975	24	27	3.9	0.6	6	14
1976	17	16	5.2	0.6	3	22
1977	16	9	5.6	0.7	10	16
1978	8	13	5.5	0.9	9	4
1979	13	15	5.1	1.0	29	7
1980	18	21	6.4	0.6	12	10
1981	12	13	10.0	0.4	4	8
1982	9	9	11.7	0.2	5	9
1983	5	9	12.4	0.5	4	10

Source: Prest, A.R. and Coppock, D.J. (1984) *The U.K. Economy*, 10th edn, p. 47, Weidenfeld and Nicolson.

familiarity of the home neighbourhood, and awareness of — and anticipation of — price increases.

Stoppages lasting less than one day were not included in 'notified' strike figures, nor were those involving less than ten workers. Probably British unions were more prone to these sudden short stoppages than unions abroad. British figures of *notified* stoppages, until 1979 at least, were lower proportionally than those of Spain, Italy, Eire, and much lower than in Canada, Finland and Australia. But they were much higher than in Japan, Holland, Western Germany and France, and significantly highly concentrated in certain industries, including mining, shipbuilding and car manufacture. So it is a surprise to learn that between 1946 and 1973 the proportion of working time lost through notified strikes was only 0.07 per cent, and that of recent years, 1.5

hours per annum have been lost per employee through strikes. The damage was probably far more profound than the figures suggest, preventing longer-term considerations by unions and employees alike, and having a bad effect on foreign buyers. When the long postwar boom showed signs of weakening in the late 1960s, and governments and journalists looked for reasons to explain why this should have affected the British economy as much as it did, much attention was given to trade unions and how they conducted their arrangements.

Trade Unions, Workers, and the Law

During this period, the Donovan Commission (1966–8) which inquired into trade unions and employee associations suggested changes which might gradually affect the legal framework within which bargaining took place. By 1969, with a good deal of public support and taking up some of Donovan's suggestions, the Labour Government produced a White Paper, 'In Place of Strife'. It proposed that the Secretary of State for Employment should have power to order a 28-day 'conciliation pause' when strikes had been called in breach of procedure, with fines for failure to observe the pause. It also proposed powers to order a secret ballot of union members before a major strike. These two clauses were bitterly resented by the unions. The proposed bill was dropped, with the TUC making a 'solemn and binding' promise to settle unofficial disputes if it could. It could not always do so.

The Conservatives passed the Industrial Relations Act of 1971, repealing the old acts of 1871, 1875 and 1906. Only unions whose objectives had been examined and approved by the Registrar of Trade Unions could be registered and called trade unions. They then had legal rights but also legal liabilities, if employers chose to take unions which broke collective agreements to court. All trade unions, registered or not, could be prosecuted for damages if their officials or 'agents', that is, shop stewards, took part in 'unfair industrial practice'. The secret ballots and 'cooling off' clauses of Labour's White Paper became law too. Their effect was very limited, because if employers used them more friction would follow. The one secret ballot held produced a vote for a (rail) strike. Other clauses delimited the closed shop (revised again in 1975 and 1980) and introduced a sensible procedure for inter-union disputes. The unions were almost totally opposed to the Act and, seeing it as an attack on cherished freedoms, refused to register. It was abolished by the Labour Government of 1974.

That government's own Trade Union and Labour Relations Act of

1974, and its 1976 amendment, was more to union liking, though it, too, brought matters which had previously been left to union discretion under the law. The Conservative Government's Employment Act of 1980 was the beginning of a new set of limitations on union activity. Some of the changes introduced, such as compulsory secret ballots for union elections and before strikes, were later accepted by the Labour Party and an increasing number of trade unions. Public money was provided for these ballots. Much more controversial were clauses allowing sequestration of union funds in certain conditions, and limiting picketing. The law had always allowed employers to dismiss strikers for breach of contract, but until the 1980s they rarely did so, since they were anxious for those same employees to resume work again.

In 1975 the Labour Government appointed Lord Bullock to head a Committee of Inquiry into Industrial Democracy, that is, worker representation on the boards of companies. Faced with a lack of enthusiasm by both employers and trade union leaders, the Labour Government did not implement the report. Recent research has shown that most employees want control only over matters which affect their immediate jobs, not over pricing policy, investments and profit distribution, the main matters for board decisions. In 1977 the newly nationalised shipbuilding and aerospace industries were obliged by law to give workers participation at board level. But these examples were not followed.

Experiments in Japanese-style management techniques — one-union representation at the workplace and 'no-strike' agreements — relied not on law, but on what some saw as union realism at a time of rising unemployment, and others as a sell-out of union rights and practices. That they have been accepted at all may owe something to a new climate of opinion fostered by successive acts, as well as disillusionment with past trade union practice. Other laws affecting unions are dealt with under Wages and Conditions of Work.

The Growth of Trade Union Membership

Figures for various sections of the labour force show that density of trade union membership in manufacturing (particularly among manual workers) was higher than in the economy as a whole. Table 12.2 shows a slight drop in density between 1948 and 1968. Between 1968 and 1979 there was an increase, particularly large among white-collar workers, previously very poorly organised, trying to restore their relative earnings. The 1974–76 Labour Government encouraged a more

Table 12.2 *Union Membership and Density by Sector in Britain, 1948–1979*

	1948		*1968*		*1979*	
	Total (000s)	*Density (%)*	*Total (000s)*	*Density (%)*	*Total (000s)*	*Density (%)*
All	9,363.0	45.2	10,200.0	44.0	13,447.0	55.4
Manufacturing	3,720.1	51.0	4,138.4	49.9	5,157.4	69.8
Manual	3,566.5	58.2	3,808.1	62.0	4,234.6	80.3
White-collar	153.6	13.2	330.3	15.4	922.8	43.7

Source: Bain, G.S. and Price, R. (1983) 'Union Growth: dimensions, determinants and density', in Bain, G.S. (ed.) *Industrial Relations in Britain*, p. 11, Basil Blackwell.

favourable attitude towards trade union membership, in spite of media criticism. In general, the larger the plant, the greater the degree of unionisation. The TGWU greatly increased its manufacturing members by merging other unions with itself, including the Chemical Workers Union in 1970 and the National Union of Vehicle Workers in 1973. White-collar unions, which included people in manufacturing, were strengthened by the merger of the Association of Scientific Workers with the Association of Supervisory Staff Executives and Technicians to form the Association of Scientific Technical and Managerial Staffs (ASTMS) in 1968. This was followed by spectacular growth, led by Clive Jenkins who listed among his interests 'mobilising the middle classes'.

Mergers between unions have not necessarily been associated with union growth. In some declining industries, once-large unions restricted to particular parts of the business have combined to form a viable entity. Growth of large diversified corporations, operating in scattered plants and different industrial sections, weakened the hold of national and industry- or craft-based trade union structures. In the 1970s 'combine committees', committees of shop stewards from different plants in a particular firm, were set up; the Lucas Aerospace Combine Committee, which drew up an 'alternative plan' for the firm, illustrated this.

Working Hours

There was no sudden reduction in the standard working hours, 1951–80. In the early 1950s about 44 hours a week was the norm; by the 1970s it was 40, usually worked in five eight-hour days. Actual working hours were on average longer because of overtime working, relied upon in many industries to produce additional earnings, though legal limits for women and children remained.

Reductions in hours were brought about by collective bargaining (and some strikes). With the increase in unemployment in the 1970s there was pressure from trade unions to reduce hours and overtime to spread work over a larger number of workers. This achieved little, since it would have added to costs. No union suggested lower wages for shorter hours.

Shift-working increased, due to the growth of continuous-process industries, as in some parts of the chemical industry, where production had to be kept going 24 hours per day. It also made best use of expensive capital equipment, though of course higher wage rates were usually paid for night work. The shorter working week and 10-hour day introduced to deal with the electricity shortage during the miners' strike of 1974 did not last. Short time was worked in 1979/80 in some industries as demand fell. However, where demand did not soon recover, dismissals and redundancies followed, and of course there were reductions in average overtime hours. Highest hours were worked by skilled men aged 25–54 in faster-growing, highly concentrated industries.

Wages

Real wages grew by more than a quarter, 1950–1965, with an increase in every year except 1962. Increases continued in the 1970s, although there was a fall in 1975, and stability in 1976. Overall, governments of both major parties welcomed, and took credit for, the growing standard of living, but also consistently attempted to slow down the rate of increase through income policies because of concern over export prices and general inflation.

There was a wide range of wage levels in manufacturing; some of the lowest were in the clothing industry, and some of the highest in car production. 'Wages drift' set in when shop stewards secured increases above the national agreements for their members. Within factories, however, differentials between grades of workers tended to get smaller, partly as a consequence of government policies, particularly in the 1970s. Women's wages were still considerably lower than men's, although the Equal Pay Act of 1970, in full force in 1975, made it illegal to pay women less than men for the same job. Other forms of discrimination remained. Women's wages for full-time work in 1980 were 71 per cent of men's, on average; part-time women workers earned 57 per cent of male hourly wages.

Table 12.3 illustrates the point made earlier about the importance of local wage agreements, particularly in all branches of engineering and vehicles.

Table 12.3 *The Most Important Levels of Pay Bargaining by Industry, 1980*

	Multi-employer		*Single employer*		
Industry	*Association*	*Wages council*	*Multi-plant*	*Single plant*	*Managerially determined*
Food, drink and tobacco	18	1	25	42	14
Chemicals, etc.	11	0	19	15	54
Metal manufacture	8	0	7 76	9	
Mechanical engineering	4	0	0	64	22
Instrument engineering	0	0	34	32	30
Electrical engineering	5	0	30	53	11
Shipbuilding	0	0	5	64	31
Vehicles	7	2	28	52	11
Metal goods not elsewhere specified	11	1	10	65	13
Textiles and leather	38	9	18	13	22
Clothing and footwear	46	14	22	2	13
Bricks, pottery, glass and cement	62	0	16	15	7
Timber and furniture	42	4	16	14	23
Paper and printing	54	1	15	17	13
Other manufacturing	8	3	31	35	22
Quarrying and construction	52	7	4	7	30
Transport & communication	22	4	39	18	17
Distribution	16	10	21	15	37
Banking and insurance	19	2	31	0	47
Education and professional services	42	4	4	2	47
Hotels and misc. services	8	24	13	7	48
Total private industry	21	5	18	30	26

Source: Bain, G.S. (ed.) (1983) *Industrial Relations in Britain*, p. 144, Basil Blackwell.

In the 1950s the idea of an annual wage increase became accepted. It was not necessarily related to productivity (in spite of the efforts of government and managers) nor, directly, to the cost of living, which it usually exceeded. Conservative governments in the 1950s endeavoured to check wage disputes by persuasion and by referring them to Courts of Inquiry. But in 1966 bad trade figures, a seamen's strike which made exports virtually impossible, rising inflation, and pressure on the pound caused a Labour government to put a legal 'freeze' on wages for six months, followed by a period of 'severe restraint'. Fines, and even imprisonment, might follow the breaking of the freeze. Exceptions and 'special cases' could be allowed, but the government could compel any suspect agreement to go to the PIB, whose findings in such cases had the force, and carried the penalties, of law.

This compulsion was completely new in modern Britain. It may well have slowed down wage claims temporarily; nevertheless, money wages rose by 5.5 per cent in the following 14 months, and disputes and strikes over wages grew. Delay in getting cases settled by the PIB only added to unrest. The difficulties of tying wage increases to measured increases in productivity, as the PIB tried to insist, became apparent. The freeze was followed in the late 1960s and 1970s by a variety of schemes, some voluntary, some compulsory, to restrain wage increases. Both political parties tried fixing maximum percentage increases, or maximum flat-sum increases for all wage and salary earners. The flat-sum increases of course reduced wage differentials. The growth in the number of shop stewards owed something to new complexities introduced by law.

It was claimed that all these interventions, statutory and otherwise, had effect only for a short time, and that wages resumed their upward course as each ended. For better or for worse, this seems too limited a view. The Labour Government's 'phases' of the 1970s did something to alter expectations. But the strife produced was in the end more destructive of industrial output than the policies were felt to be worth. When the Conservatives won the election of 1979, they abolished all compulsory wage policies, partly for this reason. They expected increases in civil servant and nationalised industry pay to be no more than in line with inflation — 13 per cent in 1979, and 18 per cent in 1980. Average weekly earnings, however, rose 15 and 21 per cent respectively, but whether the subsequent fall in both was a triumph for Toryism or a reaction to mounting unemployment or a reflection of slower growing world prices is open to debate.

Conditions of Work

There was considerable variation in the quality of working conditions. Newer plants and newer processes were light, clean and orderly; older ones were often grubby, littered with discarded equipment and carelessly stacked materials. Unions were able to negotiate from strength on these issues, although there was some resistance from members where new safeguards made work practices more difficult. Increasing awareness of the dangers of many substances — asbestos and radio-active materials — led to stricter screening of new products and processes. However, hope of employment locally led local unions to minimise possible hazards; trade unions did not speak with one voice on this. The Health and Safety at Work Act 1975 consolidated and extended previous legislation and established a statutory Health and

Safety Commission and a Health and Safety Executive to oversee compliance. Protective clothing, industrial earmuffs and protective helmets, once thought of as 'cissy', were made compulsory in a number of industries, but because the Act involved employers in heavy expense, it was slow to have full effect.

Other new laws affecting conditions of employment included the Redundancy Payments Act of 1965 which compelled employers to make payments proportionate to the length of time for which any employee with more than two years' employment for a given firm had worked, if his/her dismissal was due to the need to reduce the workforce. Laws governing 'unfair' dismissal were passed. The latter was first legislated for in the 1971 Industrial Relations Act. Paid maternity leave, better notice of dismissals and redundancies, and time off for trade union or public duties were all made law. It was not easy to bring and win unfair dismissal cases, but the redundancy laws at least cushioned the blow of dismissals in failing or rapidly automating firms. A number of major firms improved and increased their personnel departments, looking for better relations with their workforce and the unions to which they belonged.

Engineering Unions

The AEU merged with the Amalgamated Union of Foundry Workers in 1967 to create the Amalgamated Engineering and Foundry Workers Union (AEF). Further mergers in 1970 with the Constructional Engineering Union and the Draughtsmen's and Allied Technicians Association (DATA) led to the formation of the Amalgamated Union of Engineering Workers (AUEW). The union was divided into four largely autonomous sections, the Engineering Section, the Foundry Section, the Constructional Section and the white-collar Technical, Administrative and Supervisory Section (better known simply as AUEW/TASS). Only the TGWU, which of course had non-manufacturing members, was bigger.

The Procedure Agreement of 1922 remained in existence. Employers, however, still refrained from using it for fear of stoppages. By the late 1960s change was in the air. The principal issue was that, while employers could in theory impose any change at will, complaints or changes desired by the unions had to go through a long series of negotiations culminating in a central conference at York. In 1969 the union sought to compel employers to use a disputes procedure when making changes unacceptable to employees. The EEF rejected this and in 1971 the large but loosely organised CSEU pulled out of the

Agreement. The relationship between the engineering unions and the EEF was not again governed by a national agreement until 1976. The new agreement had no reference to managerial prerogatives, and there was an important 'status quo' clause, which ensured that any changes not agreed by both parties would not be imposed until conciliation had been attempted.

In the 1950s, national agreements still had a significant bearing on earnings of engineering workers, and their untions set the pace for claims by others. Engineering and shipbuilding employers fought union wage demands, but after two years had to give way after a second hearing at a Court of Inquiry. Out of this dispute came the Council for Productivity, Prices and Incomes, meant to oversee wage bargaining in industry and to take a less compromising line. Unions disliked it, though really it was not very effective. Although national bargaining subsequently became much less important in engineering as plant- and company-level bargaining grew, a national agreement in 1964 reformed certain elements of pay structure and provided a 40-hour week.

In the AEU, and later the Engineering Section of the AUEW, there were about 23,500 shop stewards in 1960, 34,000 in 1973 and 40,000 in 1980. The degree of political motivation behind their actions is open to debate, but it must be said that they made constructive national action or agreement difficult. Notable returns to national strikes were those in support of the AUEW when it was fined by the shortlived National Industrial Relations Court in the early 1970s, and the unsuccessful 1979 dispute with the EEF over pay and working hours. Hard hit by the 1979 recession, engineering employers held firm.

Cotton Textile Unions

The workforce in cotton fell from 320,000 in 1950 to 35,000 in 1982. Once a great force in Labour politics, cotton unions lost influence. In the 1950s there were considerable reforms of cotton's antique payment systems and a slow move towards shift working, often using immigrant labour. Workers received redundancy money under the 1959 Cotton Act; this set the pattern for future arrangements in other declining industries. Despite the collapse in output and the changes that took place as rayon ousted cotton, for most of this period union structure remained unchanged. Not until 1974 did the spinners and weavers unions merge to form the Amalgamated Textile Workers' Union, which in 1982 had a membership of only 23,000. In the finishing trades and the woollen industries, the principal union, the National Union of Dyers, Bleachers and Textile Workers, also suffered major membership losses. It joined the Textile and Garment Workers' Union in 1982, when it had a membership of 34,000.

Textiles provide an example of an industry with high union membership but no tradition of shop steward activity, and no major postwar strikes. The large proportion of female employment may well have been a factor in this comparatively peaceful industry.

Shipbuilding Unions

The industry's severe crisis of the late 1950s and early 1960s brought to the fore arguments about restrictive practices and demarcation, enforced by the separate trade unions, and by shop stewards at shipyard level. Labour productivity was considerably below international levels. National agreements between the Shipbuilding Employers' Federation and the CSEU set minimum and basic rates and certain conditions of employment, but actual rates of pay were determined by informal local bargaining, and the industry was characterised by wage drift. Strikes against employers, and who-does-what disputes and strikes disrupted production.

In 1962 there were attempts to overcome some of these problems. The three major shipbuilding craft societies, the Boilermakers, Shipwrights and Blacksmiths, merged to form the Amalgamated Society of Boilermakers, Shipwrights and Blacksmiths (ASB). The Shipbuilding Employers' Federation proposed the formalisation and simplification of the wages and craft structure and the 1966 Geddes Report on the industry agreed with much of this. But very little changed. The amalgamation of yards in the mid-1960s did not in itself alter industrial relations.

The collapse of the Upper Clyde Shipbuilders in 1971, and the famous (well-televised) work-in strike of that year led to some reduction in craft exclusiveness and eventual recognition among some striking workers that restrictive working practices should be abandoned to save the industry. Even so, the structure of industrial relations was not fundamentally altered until the industry was nationalised in 1977, forming British Shipbuilders. British Shipbuilders and the CSEU made formal national pay agreements and later agreed to scale down craft barriers, but only against a background of a catastrophic fall in orders for British shipyards. Govan Shipbuilders, which emerged from the collapse, insisted that its employees be trained in a variety of skills.

Within the ASB craft traditions died very slowly, and it was 1982 before it merged with the General and Municipal Workers Union to form the General, Municipal, Boilermakers and Allied Trade Union

(GMBATU). The collapse of much of the British shipbuilding industry was not solely due to union disputes and practices; however, union influence, nurtured on memories and handed-down stories of the hard times of the 1920s and 1930s, certainly contributed to its decline.

Iron and Steel Unions

The iron and steel industry had a record of harmonious industrial relations. Between 1950 and 1960 the number of steelworkers rose from 296,000 to 326,000, then fell to 255,000 in 1970. The industry was overmanned by comparison with foreign steelworks, partly because of old labour-intensive plants and machinery, but in new plants, too, there was overmanning. Management, to avoid labour disputes and meet current production targets before and after nationalisation, employed extra workers rather than insist on a smaller, more efficient and better-paid workforce.

From 1974, when demand for steel did not reach the levels assumed, there were increasing uncertainties in the workforce as old plants closed and new plants faced severe overseas competition. Unlike the shipbuilders, moderate unions in the industry accepted cuts in the workforce without strikes, helped by large redundancy payments. Union leaders sat on the industry's board after nationalisation until 1979.

By 1980 the management of the British Steel Corporation, making increasing losses as demand fell, offered its workforce a pay increase less than the rate of inflation. Rumours of further redundancies and closures as well as the pay dispute led to a national strike, the first in steel since before World War I. Private steel firms which had nothing to do with the cause of the dispute also struck and the strike lasted for thirteen weeks. It was intended to put pressure on steel users, not only on the Corporation. It was settled in a Court of Inquiry; the judgement was in effect a defeat for the union after one of the longest postwar strikes.

Until 1980, the steel industry had provided an example of strong centralised unionisation combined with peaceful industrial relations. After it, cooperation returned, spurred on by the wish to avoid closure by showing improved productivity.

Chemical Unions

In the 1960s ICI, Britain's most important chemical company, became increasingly interested in the new social sciences of management, particularly those coming out of American business schools. Some social scientists of the 1960s argued that continuous-process industries with high capital intensity, like the chemical industry, were inaugurating a new phase in workplace relations. In these industries, it was said, relations between managers and workers would become increasingly good because of the need to work together to control the expensive machinery.

ICI was not interested in these theories simply to speed seemingly inevitable social change in the workplace. It was increasingly concerned that its labour productivity was below that of foreign competitors. It thus launched into a major reform of working practices and payment systems in 1965, starting with extensive trials of schemes to break down job demarcation and redesign jobs, known by the acronym 'MUPS' (Manpower Utilisation and Payment Structure). These met resistance, but agreement was reached after long negotiations with unions' general secretaries and national officers on higher pay, a simpler wage structure (getting rid of many bonus systems), and improved working conditions. The Weekly Staff Agreement of 1969, unlike many of the so-called productivity agreements of the 1960s, was genuine. It also promised future change.

ICI's productivity increased, but continued to lag behind competitors. Management argued that it was still overmanned and the most severe cuts in its UK workforce came after 1976, though some reduction was due to closure or sale of whole plants rather than to reductions in manning levels. That the firm was not more handicapped by overmanning than it was must have been due to (1) the relatively small percentage of labour costs in total costs; (2) the lower wages paid in Britain compared with those paid to workpeople in the USA and Germany; and (3) the general efficiency of the company.

Motor Vehicle Unions

In the 1970s the motor industry seemed to many to sum up the problems of British industry. It became without doubt one of the most strike-prone in British manufacturing, and by the 1970s had lower productivity and a lower general quality of output than its competitors in Europe and Japan. In the 1950s and 1960s the industry had expanded rapidly; the wages it paid were above average for British manufacturing. Work was particularly noisy and monotonous, however; its availability varied from month to month and year to year, and workpeople were 'laid off' with little or no notice.

Trade union organisation grew, particularly in the 1960s, and by the early 1970s all the major manufacturers were 100 per cent unionised. The strength of shop stewards increased, especially in the British-owned firms where, until the 1970s, production workers were paid piece rates argued over by shop stewards. Many of the changes made by management in the 1970s were directed towards reducing or eliminating shop steward involvement in pay and manning decisions.

One of the first attempted reforms when BLMC (eventually called Austin Rover) was formed was the replacement of piece rates by Measured Day Work. This system was closer to a day wage, but involved a bonus element calculated on a plant rather than an individual or group basis. Its introduction led to many disputes. The firm's crisis and its nationalisation in 1975 led to adoption of a recovery plan which included formal consultation with shop stewards about most aspects of the plan, and also further moves towards the centralisation of pay bargaining.

The appointment of Sir Michael Edwardes to the chairmanship of BL in 1978 signalled a determined challenge to shop stewards. He went over their heads, and balloted workers directly on proposals for changing working practices without which British Leyland would be closed. Edwardes' plan was approved by an 87 per cent vote in favour, despite opposition from many shop stewards. An eventual outcome was the dismissal of the convenor of shop stewards at the Longbridge plant, who was also chairman of the firm's combine committee. During his 33 months of office, only 4 per cent of days had been free of disputes.

Between 1980 and 1984, BL experienced a 130 per cent improvement in productivity. Forty per cent of this improvement was said to be due to new, more automated technology. But over 50 per cent was said to be due to new agreements on working practice and manning levels and an uninterrupted flow of work.

In 1955 Ford signed a Procedure Agreement with 22 unions with members in Ford. Each union had a single vote on Ford's National Joint Negotiating Committee, to which all matters that could not be settled locally would be referred. The FNJNC formalised Ford's principle of dealing with national union officials, rather than shop stewards or district representatives. Nevertheless, Ford was forced, although reluctantly, to recognise shop stewards. The relationship was extremely uneasy, however, and in 1962 there was a 'purge' of 17 shop stewards. This led to a strike at Dagenham and the establishment of a Court of Inquiry. Not even Ford could dictate terms unilaterally.

When Ford built a major new site at Halewood on Merseyside it wished to deal with only a few unions, and to prevent shop steward activity. It signed a recognition agreement with only the AEU and the

NUGMW. The TGWU threatened to black Ford cars at Liverpool docks, and so it, too, was recognised, and in fact became the largest union at Halewood. Furthermore, contrary to Ford's wishes, shop steward organisation grew quickly.

In 1969, the publication of *In Place of Strife*, and the statutory incomes policy which made it illegal to grant pay rises above a certain level without productivity increases, gave Ford an opportunity. It signed an agreement with the representatives of the 22 unions on the FNJNC. In return for a pay increase of 7.5 per cent, they accepted penalty clauses under which workers would lose pay if they took 'unconstitutional' action, that is, an unofficial strike.

But Ford workers did not accept it. Unofficial strikes began at Halewood, and were quickly made official by the AEF and TGWU. The new radical leadership of these two unions backed the strikes, which soon covered all Ford's UK plants. After three weeks the government arranged an inquiry and the strikes were settled; the penalty clauses never came into operation. A consequence was the inclusion of convenors in future negotiations. In 1978 Ford was again involved in a strike of major political significance over pay, one of the largest strikes in the car industry in the 1970s. The outcome was a pay increase (which it could afford) and the breaking of the Labour Government's incomes policy.

That Ford management, in spite of its resistance, nevertheless had been compelled to agree to a degree of shop steward and convenor control of the work process is shown in the proposals put to the workforce in early 1986. These included cutting down the number of different job descriptions at the Dagenham works from over 500 to just over 50. The remarkable thing is that over 500 job definitions with their

Table 12.4 *Strike Activity in Motor Vehicles, 1949–83*

Period	*Average no. of strikes p.a.*	*Average working days lost p.a. (000s)*[a]
1949–53	25.4	170.4
54–58	35.4	125.6
59–63	106.2	377.6
64–68	185.6	544.0
69–73	272.6	1,853.2
74–78	194.0	1,893.6
79–83	105.2	1,069.0

Source: Marsden, D. *et al.* (1985) *The Car Industry: Labour Relations and Industrial Adjustment*, p. 121, Tavistock.

Note: [a] By workers directly involved.

associated rigidity and wage agreements had been accepted by a reputedly 'strong' management in the first place.

Conclusion

Much of this chapter has dealt with conflict. It is the aspect of union affairs most often given publicity, and has been the most frequent cause of government intervention in the relationship between management and workforce.

To the followers of Marx, these conflicts are inherent in the structure of our society, to be resolved only by the forcible overthrow of the existing class structure and its replacement by a new order, admittedly not yet fully in being anywhere. To the less dogmatic historian, the memories of disputes and the grim working conditions of the past stand in the way of reasonable solutions to present problems. Some of the newer initiatives in management skills and in worker participation in decision making might be able to reduce these shadows. Traditional adversarial attitudes are no help to any industry. What is left out of a picture which focuses on conflict is the shared interest of management and worker alike in the production of a saleable object.

Further Reading

Coates, K. and Topham, T. (1980) *Trade Unions in Britain*, Spokesman Books.

Fox, A. (1985) *History and Heritage: The Social Origins of the British Industrial Relations System*, Allen and Unwin.

Pagnamenta, P. and Overy, R. (1984) *All Our Working Lives*, British Broadcasting Corporation.

Questions

(1) What were the similarities and differences between trade unions in shipbuilding and in iron and steel, 1951–80?

(2) Why were industrial relations so poor for so much of the time in the car industry, 1951–80?

(3) How important were changes in the law affecting trade unions, 1951–80?

(4) Would the Bullock Report's recommendation of worker representation on the board improve employer/employee relations if adopted in Britain?

Summing-Up

The years 1870–1914 saw British manufacturing output expand, yet face challenges in world markets, and to some degree at home. Older industries responded to uncertainty by restraining capital expenditure, relying on the undoubted skills of the British workforce to produce goods of quality. Those skills, however, could not compensate for the lack of scientific training needed for leadership in the new industries. The workforce was understandably wary of innovations which might use less of their skills.

From 1914 to 1939, much of Britain's economic development was affected by World War I, which reinforced demand for the products of the older industries, as did the brief period of prosperity which followed. Then the downturn in world trade, accentuated by the 1929–33 Depression and caused in part by the drive for self-sufficiency in many countries, hit British exports. Old industries lost markets and the consequent unemployment soured industrial relations for decades. New industries, while not technically innovative, grew in a protected home market, and even found some export markets. Employers still restrained capital outlay, unable to foresee an expansion of sales which would be bound to cover the outlay. Successive governments, partly because of the tradition of non-intervention, and perhaps partly because of a reluctance to increase unemployment in the short term, were slow to encourage re-equipment. Yet, while the Depression was severe in Britain, its impact was not as great as in Germany or the USA. Trade union pressure kept up real wages in Britain, and so helped to prevent the home market from collapsing to the same extent as in those countries.

In the time span 1939–51, World War II strengthened demand in Britain and elsewhere: applied science assumed a new importance in weaponry, where cost was a secondary consideration. By 1941, the prewar economic situation was completely reversed in Britain. There was a shortage of manpower, of raw materials, and of factory space. Every manufactured item was sold, often to the government. Postwar, sales continued, but with government purchases scaled down and the home market restricted in order to encourage exports. Manufactured exports were urgently needed to pay for peacetime imports: the war had cut down income from abroad, and the sterling reserves, although helped by commonwealth and empire holders, were not strong enough to withstand sudden runs or shocks. Fortunately, the world was willing and able to absorb British exports. European and Japanese producers had been much damaged in the war and British goods, while produced in an atmosphere of shortage and constraint, found markets with comparative ease. The coal industry had been at the centre of industrial unrest during the war: in manufacturing there had been rather more willingness to work together for a common purpose, and this lasted into the peace.

The years 1951–80 at first saw continuing growth and prosperity in British manufacturing and the British economy, with the USA the admired model of efficiency, productivity, and high wages. Unable to compete in scale of output, British manufacturers nevertheless continued to find markets without making profound changes in their mode of production. But in the later 1960s and 1970s it became apparent that economies which had begun from a lower starting point, and might therefore be forgiven for growing faster than the British, were making inroads into British markets at home and abroad, while paying their workforce more. They had greater productivity than British industries, although the extent to which this was due to more efficient work practices or higher capital input differed from industry to industry, and was much open to debate. As the world became more prosperous, design and performance of goods became more important and here Britain lagged behind. Strikes, inflation and the oil price rise of 1973 added to British difficulties. Some industries, such as the chemical industry, took decisive action to put their house in order, but many companies and industries were in trouble. Governments tried to help in a number of ways, and did succeed in nursing some firms through the worst times. These came when the increased value of the pound (due to the 1979 oil price rise) made it harder to export, and easier still to sell imported goods in Britain. By 1980 it was accepted by employer and employee that a redesign of the product and the process of production was essential to the survival of firms in a competitive world in which low-wage countries had sophisticated technologies. The industries even of the mighty USA were under challenge from the Far East.

For two hundred years, manufacturing industry had been an engine of great power in the British economy. A cluster of services, some physical, some financial, had grown up around it, and much of British employment had depended upon it, directly and indirectly. If, to achieve the same or greater output with less cost or in less time than formerly, fewer people were to be employed in manufacturing, that did not lessen its value as an earner of foreign currency, or an employer of services, or a forestaller of imports. North Sea oil diminished the need for some of these roles, but did not obliterate them. It was more important than ever to have well-trained people to attend to R & D, to have adequate investment, and to have managers and unions, employers and employees, who saw the need for cooperation across the great divides of class consciousness and memories of past hostilities.

Many factories and workplaces have become almost unrecognisably cleaner and quieter and safer than those of a hundred years ago. Further change in the production and use of goods may come about through increasing concern for the environment of Britain and worldwide efforts to prevent pollution and climatic shifts. Wage differences between countries will matter less if manufacturing industry becomes more automated: transport costs will then become more important in affecting the final cost of goods in any given market. This book ends, but the story of manufacturing industry in the British economy continues.

Statistical Appendixes

1. UK Manufacturing, 1856–1973: Growth of Labour, Capital, Total Factor Input, Output, Total Factor Productivity, and Total Factor Productivity Weighted by the Share of Manufacturing in the UK GDP, 1856–1973 (annual percentage growth)

	1856–73	*1873–1913*	*1924–37*	*1937–51*	*1951–64*	*1964–73*
Labour	1.1	0.8	1.4	1.0	0.2	−1.6
Capital	3.2	2.6	1.0	2.9	3.2	3.3
Total Factor Input	1.7	1.4	1.3	1.6	1.2	−0.1
Output	2.6	2.0	3.2	2.5	3.2	3.0
Total Factor Productivity	0.9	0.6	1.9	0.9	2.0	3.1
Total Factor Productivity, Weighted by Share of GDP	0.20	0.16	0.58	0.29	0.70	1.07

Source: Matthews, R.C.O., Feinstein, C.H. and Odling-Smee, J.C. (1982) *British Economic Growth 1856–1973*, Table 8.3, Oxford University Press.

Note: Land, which together with labour and capital forms the classic trio of factors of productivity, is omitted from this table; it is a relatively minor cost in manufacturing. Capital is used here in its widest sense of 'all those man-made aids to further production such as tools, machinery, plant and equipment which is not consumed for its own sake, but is used in the process of production of other goods and services' (Lipsey, Richard G., 1979, *An Introduction to Positive Economics*, 5th edn, p.52, Weidenfeld and Nicolson.

II. The United Kingdom's Share of World Exports of Manufactures Compared with Six Other Industrial Countries, 1881–1973 (Percent, based on values in US $ at current prices)

Year	*UK*	*US*	*Germany*[a]	*Italy*	*France*	*Japan*	*Sweden*
1881–1985[b]	43.0	6.0	16.0	2.0	15.0	0.0	1.0
1899	34.5	12.1	16.6	3.8	14.9	1.6	0.9
1913	31.8	13.7	19.9	3.5	12.8	2.5	1.5
1929	23.8	21.7	15.5	3.9	11.6	4.1	1.8
1937	22.3	20.5	16.5	3.7	6.2	7.4	2.8
1950	24.6	26.6	7.0	3.6	9.6	3.4	2.8
1964	14.0	20.1	19.5	6.2	8.5	8.3	3.4
1973	9.1	15.1	22.3	6.7	9.3	13.1	3.3

Source: Matthews, R.C.O., Feinstein, C.H. and Odling-Smee, J.C. (1982) *British Economic Growth 1856–1973*, p. 435, The Clarendon Press.

Note: Total for world exports excludes exports from small manufacturing countries and the postwar Soviet bloc.

[a] For 1881–1937 estimates are 71% of contemporary Germany.

[b] The comparability between the estimates for 1881–85 and those for later years is imperfect.

III. United Kingdom Trade in Manufactured Goods: Ratio of Export Value to Import Value (100 = equilibrium)

	1963	*1968*	*1973*	*1978*	*1979*	*1980*	*1981*	*1982*	*1983*	*1984*
World	219	143	115	115	104	112	109	101	89	
EEC	164	121	88	83	80	89	81	74	66	
USA	98	95	110	94	85	70	76	75	84	
Japan	149	107	53	35	33	29	23	21	18	
Other OECD	255	148	106	94	92	106	100	96	88	
Oil Exporters	x	484	x	x	x	926	x	x	762	
Latin America	310	286	314	356	269	253	303	225	146	
Comecon (Europe)	223	146	155	189	170	217	196	154	150	
Asean	x	x	337	319	224	253	209	229	185	
African	565	187	223	467	x	x	x	x	655	

Source: Report from the Select Committee on Overseas Trade (1985) Vol. I, p.20, HMSO.

Index

Industrial Studies Series

Series Editor: J F Pickering

The Firm and its Environment
R H Barback

Technology and Production
G Clews and R Leonard

Manufacturing Industry since 1870
Margaret Ackrill

Contemporary Trade
F N Burton

Technological Development
Dale Littler

Marketing and Product Development
Dale Littler

Management Structures and Techniques
Roderick Martin and Brian Moores

Employment Relations in Industrial Society
John Goodman

Structure and Performance of Industries
T T Jones and T A J Cockerill